How to Save Our Planet

The Facts

PROFESSOR MARK MASLIN

PENGUIN LIFE

AN IMPRINT OF

PENGUIN BOOKS

PENGUIN LIFE

UK | USA | Canada | Ireland | Australia
India | New Zealand | South Africa

Penguin Life is part of the Penguin Random House group of companies
whose addresses can be found at global.penguinrandomhouse.com

First published 2021

003

Set in 11/15 pt FS Industrie Cd
Typeset by Jouve (UK), Milton Keynes
Printed and bound in Great Britain by Clays Ltd, Elcograf S.p.A.

The authorized representative in the EEA is Penguin Random House Ireland,
Morrison Chambers, 32 Nassau Street, Dublin D02 YH68

A CIP catalogue record for this book is available from the British Library

ISBN: 978-0-241-47252-1

www.greenpenguin.co.uk

How to Save Our Planet

Mark Maslin is a Professor of Earth System Science at University College London. He is the former Director of the UCL Environment Institute and a leading voice in the battle against climate change.

Mark's areas of scientific expertise include understanding the origins of the Anthropocene, the causes of past and future climate change, and the environmental challenges facing humanity in the 21st century. Mark is a Royal Society Wolfson Research Scholar and has been a Royal Society Industrial Fellow. He has published over 175 papers in journals such as *Science*, *Nature* and *The Lancet*.

Mark is the author of eight popular books including *The Human Planet* and *Cradle of Humanity*. He regularly speaks at major literary and science festivals, has written articles for publications like *The Times*, *New Scientist*, *New York Times*, *Independent* and *Guardian*, and has appeared on *Newsnight*, *Time Team*, *BBC News* and *Dispatches*. He was also an advisor and key contributor to David Attenborough's 2019 BBC One programme *Climate Change: The Facts*. Mark is an independent advisor to the Corporate Responsibility Board at Sopra Steria.

To my mother,
Catherine Anne Maslin
(1943–2020).

Without whom I would not
be the person I am today.

For everyone who is prepared
to fight for a better world.

CONTENTS

Facts and figures about the origin of the Universe and Earth, our unique planet. This chapter provides you with background knowledge of our precious planet, the evolution of life on Earth and the emergence of modern humans.

Facts and figures about the development of human society. This chapter provides you with background knowledge of the whole of human history from the hunter-gatherers through to the reign of consumer capitalism.

CHAPTER 3: STATE OF OUR WORLD   46

Facts and figures about the huge human impact on Earth. This chapter provides you with insights into the state of our planet, both in terms of the environment and our global human society.

CHAPTER 4: TAXONOMY OF DENIAL   64

These are confusing times, with a lot of misinformation and fake news about climate change and environmental destruction. This chapter provides you with an insight into the tricks used by climate/environmental change deniers to dissuade you from taking rapid action.

CHAPTER 5: POTENTIAL FUTURES – NIGHTMARE OR ECOTOPIA?   76

This chapter provides you with a clear vision of how bad the world could be if we do not deal with climate change and environmental degradation. It also provides an alternative vision of what the world could be like if we do everything possible to alleviate our climate, environmental and social crises.

CHAPTER 6: POWER OF THE INDIVIDUAL

We have the power to control our own lives and influence those around us – this chapter provides a clear set of positive actions that you as an individual can take. Individuals are our powerbase for change.

CHAPTER 7: CORPORATE POSITIVE POWER

Businesses control our lives: what we eat, what we buy, what we watch and even who we vote for. This chapter discusses how we can harness the positive power of corporations, so we can change the world. Companies are one of our major weapons of change.

CHAPTER 8: GOVERNMENT SOLUTIONS

Governments look after our safety and our wellbeing. They control the aspirations of civil society through the rule of law and the development of policy. This chapter discusses how regulation, taxation, subsidies and incentives all play a role in how governments can make our societies more sustainable and ultimately reduce our carbon emissions to zero. Governments drive innovation.

PREFACE

The global awareness of climate change and the environmental crisis has grown very rapidly – through Greta Thunberg-led schools strikes, Extinction Rebellion, the recent Intergovernmental Panel on Climate Change (IPCC) high-impact reports, the BBC One documentary *Climate Change: The Facts* and governments all around the world declaring we are in a climate emergency.

This awareness is continuing to grow.

The science shows us that our planet and our species are facing a massive crisis, which we have caused.

Such is its urgency that we need to act now.

There are many books out there on climate change.

Some want to scare you and some want to preach at you.

This book is different.

It is for anyone and everyone to pick up and read.

I wanted to write a book that makes people feel smarter, more knowledgeable and empowered to act.

This is a book you can quote in the pub or at a dinner party or even in Parliament.

This is a new style of book to engage everyone who wants to make a difference.

This is a short and punchy handbook that will empower you by providing the knowledge and insight to act to save our planet.

* * *

This is not a linear book.

You do not have to start at the beginning or the end.

Each chapter is self-contained.

Choose the most relevant chapter for you and read that one first.

The book is inspired by Sun Tzu's *The Art of War*. Like that book, it is written in short sentences and phrases – each one laden with meaning.

Because to save the planet and ourselves we need to be on a war footing – we need to engage every part of our society in the battle against climate change and environment destruction.

Each factual statement is supported by numbered key references that are listed in full at the end of the book, so that you can check every statement I make.

I want this book to be read by everyone – because I honestly believe that together we can make this a better, safer, healthier planet for all humanity.

INTRODUCTION

WE STAND AT THE PRECIPICE.

THE FUTURE OF OUR PLANET IS IN OUR HANDS.

In the second half of the 20th century, our creativity and co-operation brought about unprecedented peace, prosperity and individual freedom.

Yet our institutions and the paradigms that enabled these successes are not keeping pace with our rapidly changing world.

Growing environmental stress, climate change, accelerating technological change and intensifying social inequality increasingly appear beyond the ability of our governments and our societies to manage.

We have entered a new geological period of time – the Anthropocene.

We have become the new geological superpower on our planet – changing the climate and the environment quicker than the moving of the continents or the waxing and waning of the great ice ages.

Climate change is the greatest threat we have ever faced.

WE HAVE ENTERED A NEW GEOLOGICAL PERIOD OF TIME – THE ANTHROPOCENE

FACTS HAVE THE POWER TO CHANGE OUR WORLD

Climate change denial is still rampant, paralysing all areas of our society.

We can all help to deal with denial – this very human emotion.

Facts are power.

Facts have the power to change our world.

We need to develop new modes of thinking for the 21st century to creatively and collectively tackle these challenges.

This book provides a simple, powerful, indisputable message about our future and the world we decide to leave our children.

No waffle, just facts.

This is a clear call for action.

The 21st century is when we save our planet from ourselves.

CHAPTER 1:
HISTORY OF OUR PLANET

THE PAST IS THE WINDOW THROUGH WHICH WE MUST VIEW OUR FUTURE.

Our Universe is 13.8 billion years old.[1]

In the beginning there was the Big Bang, when all the matter in the Universe was created and exploded outwards.[2, 3]

380,000 years of expansion cooled the newly created matter, allowing negatively charged electrons to be trapped by positively charged protons, forming hydrogen gas.[4]

There were no galaxies, no stars, no planets, no people, no life – only an expanding cloud of gas.

Our Universe contains 5% ordinary matter and energy, 27% dark matter and 68% dark energy.[1]

Gravity pulled on the irregular cloud of gases, creating clumps that formed galaxies.[5]

Within each galaxy, gravity pulled matter together to form billions and billions of stars.[6]

The immense gravitational pressure in each star forced hydrogen atoms to collide and fuse, creating helium and releasing huge amounts of energy: making the stars shine brightly in the night sky.[7]

As each star continues to burn, the hydrogen and helium continue to collide and fuse creating heavier elements such as carbon.[8]

These heavier elements collide and fuse and create the following elements: lithium, beryllium, boron, carbon, nitrogen, oxygen, fluorine, neon, sodium, magnesium, aluminium, silicon, phosphorus, sulphur, chlorine, argon, potassium, calcium, scandium, titanium, vanadium, chromium, manganese and iron.[8]

Elements from hydrogen to iron on the periodic table are referred to as the planet-building elements.[9]

Elements heavier than iron require the input of considerable amounts of energy to form. They are not created in a star but are made when a massive star (10 to 25 times the size of our Sun) explodes as a supernova.[10]

We are all made of stardust.

Our Solar System was created by the gravitational collapse of a giant interstellar gas cloud 4.568 billion years ago.[11]

Our Sun contains 99.86% of the mass of the Solar System. 90% of the remaining mass is contained in the great gas giants Jupiter and Saturn.[12]

Our home planet, Earth, is 149.6 million km from the Sun: the optimum distance (making it not too hot and not too cold) to allow liquid water to exist at the surface.[13]

Our planet has enough water and carbon at the surface to support life.[14]

Our Moon orbits very close to the Earth and acts as the Earth's gyroscope, stabilizing the rotation of our planet.[15]

Life first appeared on Earth, in the form of bacteria, between 3.77 and 4.28 billion years ago.[16]

37% of your DNA comes from these bacteria.[17]

More complex 'eukaryotic' cells, with nuclei and other internal structures, emerged between 2.1 and 1.6 billion years ago.[18]

28% of your DNA comes from these early eukaryotes.[17]

Complex life took another billion years to evolve and appeared about 600 million years ago.[19]

37% OF YOUR DNA COMES FROM ANCIENT BACTERIA

Major animal groups recognized today did not appear until the rapid diversification known as the Cambrian explosion that started about 541 million years ago.[20]

16% of your DNA comes from these weird and strange early creatures that evolved in the ancient tropical shallow seas.[17]

Vertebrates originated about 525 million years ago. Your earliest backboned ancestors were sea-floor filter feeders – sifting food out of mud on the bottom of the ancient tropical shallow seas.[21]

13% of your DNA comes from these worm-like early vertebrates.[17]

The earliest mammals appeared about 225 million years ago, near the end of the Triassic Period.[22]

Mammals are defined by having milk-producing glands in females, fur or hair, three bones in the inner ear, and a neocortex.[23]

The neocortex is the region of the brain that controls higher functions such as sensory perception, generation of motor commands and spatial reasoning. In humans, it enables conscious thought and language.[24]

For 170 million years the dinosaurs ruled our planet. Our ancestors (rodents and small, nocturnal mammals weighing as little as 2 grams) remained in their shadow.[25]

The domination of the non-avian dinosaurs ended when they went extinct 66 million years ago due to a long period of global climate change and a meteorite impact.[26]

Volcanic eruptions across the whole of India lasting nearly a million years pumped billions of tonnes of carbon dioxide into the atmosphere, warming up the Earth and causing the first dinosaur extinctions.

The final blow was the devastating impact of a meteorite in Central America, which destroyed our planet's surface and left it devoid of plant life.[26]

Our mammal ancestors survived by eating insects and aquatic plants. Many people to this day have the enzyme, called chitinase, that allows the digestion of insects.[27]

Mammals evolved to fill many of the ecological niches left by the dinosaurs.[28]

True primates and social monkeys evolved 55 million years ago, during a period of super global warming called the Palaeocene–Eocene Thermal Maximum.[29]

Living in larger groups meant that individuals had to negotiate complex webs of friendships, hierarchies and rivalries – as we do today.[30]

6% of your DNA comes from these first primates.[17]

Between 10 and 5 million years ago our ancestors evolved the ability to efficiently move upright on two legs.[31-33]

Our upright walking ancestors started using stone tools approximately 3.3 million years ago.[34]

Rapid climate change in Africa 1.8 million years ago forced the evolution of hominin species, including *Homo erectus*, with brains up to 80% larger than their ancestors.[35]

Homo erectus was our first ancestor to disperse out of Africa and spread into Asia, Europe and Australasia.[36]

The large brain of *Homo erectus* was accompanied by other significant changes to life history:

> shortened intervals between births,
> delayed child development,
> shape of the pelvis to give birth to a larger head,
> shoulder morphology that allowed projectile use,

adaptations to long-distance running,
and social behaviour.[37-43]

Between 1.8 and 1.0 million years ago our ancestors domesticated fire.[44]

Our species, *Homo sapiens*, emerged 300,000 years ago in East
Africa and dispersed into Eurasia.[45]

150,000 years ago there is the first evidence of symbolic behaviour, with discoveries of shell engravings, ochre and shell beads.[46]

60,000 years ago *Homo sapiens* 2.0, or modern humans, emerge
in East Africa, with consistent evidence of creative thinking
expressed through art, ornaments and symbolic behaviour.[47]

Creative thinking steadily increased in complexity and frequency with more and more knowledge being generated and
passed on to the next generation.[48]

Modern humans spread to every continent in the world.[49]

Our cumulative culture allows knowledge to grow with every
generation.[50]

This process continues to accelerate and has enabled humans
to become the world's apex predator and take over the planet.[51]

NOW WE ARE CREATING GLOBAL CLIMATE CHANGE

Climate change in Africa created modern humans.[23]

Now we are creating global climate change.[51]

This climate change is threatening to destroy us.[51]

We need to use all our evolutionary gifts to save ourselves and our planet.

CHAPTER 2:
HISTORY OF HUMANITY

HISTORY TEACHES US LESSONS WE DO NOT WANT TO LEARN – BUT WE MUST IF WE ARE TO SURVIVE THIS CENTURY.

In the beginning we evolved in Africa and then spread to the four corners of the Earth.[1]

HUNTER-GATHERING

Homo sapiens 2.0 swept across the globe.[1]

Expanding from Africa into Europe, Asia and Australasia.[1]

12,000 years ago we spread across the Bering Sea into the Americas.[2]

At the end of the last ice age, we reached all of the major land masses on Earth except Antarctica.[3]

In each new land we slaughtered all the large animals, the megafauna.[4]

By 10,000 years ago we had caused the extinction of 4% of all mammal species.[4]

Africa lost 18%, Eurasia 36%, North America 72%, South America 83% and Australia 88% of their large-bodied mammals.[4]

This slaughter was done by a human population of less than 10 million.[5]

Confirming our status as the world's apex predator.[6]

We need about 120 watts of power when resting, equivalent to the power required by 2 old-fashioned lightbulbs.[7]

Our hunter-gatherer ancestors used the equivalent of about 6 lightbulbs, around 300 watts per day of energy, for getting and preparing food and making and maintaining fire.[7]

AGRICULTURALISM

The end of the last ice age and the extinction of the megafauna drove us to domesticate plants and animals.[8]

Agriculture appeared independently in at least 14 places around the world:[9]

> 10,500 years ago in south-west Asia, South America and central East Asia.[9]

7,000 years ago along the Yangtze and Yellow Rivers in China and in Central America.[9]

5,000 years ago in the savanna regions of Africa, India, South East Asia and North America.[9]

Early farmers created the first energy revolution by generating 2,000 watts of power, the equivalent of just over 30 lightbulbs, through farming.[7]

Farmers spread across the world, pushing hunter-gatherers into the marginal lands, unfit for agriculture.[8]

We started to live in villages, towns and cities.

Living closely with our domesticated animals created new infectious diseases.[10]

The 25 major human diseases can be split into two types – 'tropical' and 'temperate'.[10]

The 10 major tropical diseases include: Chagas disease, cholera, dengue haemorrhagic fever, East and West African sleeping sicknesses, *falciparum* and *vivax* malaria, .tropical yellow fever, AIDS and visceral leishmaniasis.[10]

These diseases are typically transmitted by insects and cause long-term effects lasting months to decades. They evolved over millions of years with humans in Africa.[10]

The 15 major temperate diseases include: hepatitis B, influenza A, measles, mumps, pertussis, plague, rotavirus A, rubella, smallpox, syphilis, temperate diphtheria, tetanus, tuberculosis, typhoid and typhus.[10]

These temperate diseases have recently been joined by Asian flu (1956), SARS (2002), H7N9 (2013) and Covid-19 (2019/20).

These diseases either kill you or you survive with long-lasting immunity.[10]

These diseases create epidemics in dense populations of humans.[10]

Early agriculture was so successful it led to major empires rising and falling on each continent.[8]

The deforestation of land for farming releases carbon dioxide into the atmosphere.[11]

Early farmers caused atmospheric carbon dioxide to rise from 260 ppm (parts per million) some 7,000 years ago to 280 ppm by the beginning of the Industrial Revolution in the 18th century.[11]

A rise of 0.003 ppm per year.[11]

Wetland rice and ruminants such as cattle, sheep and goats produce vast quantities of methane.[12]

Early farmers caused atmospheric methane to rise from 580 ppb (parts per billion) some 5,000 years ago to 720 ppb by the beginning of the Industrial Revolution.[12]

A rise of 0.03 ppb per year.[12]

Early farmers increased these two powerful greenhouse gases in the atmosphere just enough to stop the next ice age starting.[13]

By the 16th century the human population had risen to between 425 and 540 million.[14]

MERCANTILE CAPITALISM

The geography of Europe created a multitude of small independent kingdoms which were constantly at war.[8]

This intense competition and rivalry in Europe drove Europeans to search for new lands to conquer and colonize.[8]

WITHIN 100 YEARS 56 MILLION INDIGENOUS PEOPLE IN THE AMERICAS WERE DEAD

European first contact with the Americas occurred in 1492 when Christopher Columbus landed in the Bahamas.[15]

Within 100 years 56 million indigenous people in the Americas were dead.[16]

90% of the indigenous population, representing 10% of the global population, died.[16]

This 'Great Dying' is one of the largest ever human mortality events in proportion to the global population.[16]

Only the Second World War killed more people.

Indigenous Americans had never been in contact with the Eurasian pathogens and had no natural immunity to influenza, smallpox or measles.[17-19]

Warfare, famine and colonial atrocities completed the slaughter.[20, 21]

European arrival in the Americas started the first global trade circuits.[22]

Manufactured goods from Europe, such as cloth and copper, were traded in Africa for African slaves.[22]

90% OF THE INDIGENOUS POPULATION, REPRESENTING 10% OF THE GLOBAL POPULATION, DIED

The slaves were transported to the Americas to produce cotton and 'drug foods', such as sugar and tobacco, which were sold back to Europe.[22]

Mexican and Bolivian silver was exported to Spanish-held Manila to be traded with Chinese merchants for silks, porcelain and other luxury goods, which were sold in Europe.[22]

So started the global mixing of humans, plants and animals which continues to today.[22]

The Americas gained: wheat, rice, sugar, bananas, horses, pigs, cattle, goats, chickens, epidemic diseases and slaves.[22]

Eurasia and Africa gained: maize, cassava, potatoes, sweet potatoes, beans, peanuts, squashes, pumpkins, tomatoes, chilli peppers, avocados, pineapples, cocoa and tobacco.[22]

European expansionism also created a new renaissance of the sciences.[23]

In 1543 Nicolaus Copernicus in his book *On the Revolutions of the Celestial Spheres* showed us that the Earth revolves around the Sun.

In 1620 Francis Bacon in his book *The New Instrument of Science* stated 'knowledge is power'.

In 1687 Isaac Newton published his *Philosophiæ Naturalis Principia Mathematica*.

Mercantile capitalism was founded on investing in overseas wealth extraction and created competition among ruling elites, leading to intense asset stripping of colonized countries.[24]

Over 10 million slaves were imported into the Americas to grow export crops and work in the silver mines.[25]

The first global mega-corporations were created.[26]

The Dutch East India Company colonized present-day Indonesia and by 1669 had 150 merchant ships, 40 warships, 50,000 employees and a private army of 10,000 soldiers.[27]

The British East India Company colonized present-day India, Bangladesh and Pakistan. At the height of its rule in India, in 1803, the company had a private army of about 260, 000 – twice the size of the British Army at that time.[28]

These companies controlled whole regions of the world; they could put down rebellions, imprison and execute prisoners, and essentially do whatever they deemed acceptable to extract profits.[26]

INDUSTRIAL CAPITALISM

In the second half of the 18th century the Industrial Revolution occurred in one place – Britain.[29]

Within 50 years it had spread to the whole of Europe, North America and Japan. It is still spreading today.[29]

This revolution was driven by:

A significant rise in the population of Britain between the 16th and 18th centuries, due to the huge rise in British agricultural productivity and importation of food from the colonies.[30]

The development of a large urban working class, as fewer people were needed to work the land.[30]

No internal trade barriers within Britain to hinder commerce.[3]

Britain had an ample supply of naturally occurring coal to replace the water-driven and wood-burning mills and factories.[30]

British colonial exploitation created a large wealthy capitalist class.[30]

The constitutional monarchy and strong state enforcement of laws and property rights allowed venture capitalists to fund potentially profitable technological innovations.[3]

The Victorian super-rich funded science as never before with new discoveries and inventions made in every conceivable field, from new medicines to the theory of evolution, from the invention of photography to the development of the railways.[30]

In 1804 the world population rose for the first time to 1 billion people.[31]

Just 123 years later in 1927 it had doubled to 2 billion people.[31]

Population increase occurred because communicable diseases were brought under control through sanitation and the supply of safe drinking water.[30]

The Second Industrial Revolution occurred from 1870 onwards with the expansion of railways and the telegraph, large-scale steel and iron production, use of steam power, petroleum and the beginnings of electrification.[30]

The Industrial Revolution led to the age of pollution – with waste materials being dumped into rivers, lakes, soil, oceans and the atmosphere.[32]

IN 1804 THE WORLD POPULATION ROSE FOR THE FIRST TIME TO 1 BILLION PEOPLE

JUST 123
YEARS LATER
IN 1927 IT
HAD DOUBLED TO 2
BILLION PEOPLE

Carbon dioxide levels rose from 280 ppm (parts per million) at the beginning of the Industrial Revolution to 310 ppm by the outbreak of the Second World War.[33]

A rise of 0.3 ppm per year, 100 times faster than in the previous 5,000 years.[33]

By the end of this Industrial Revolution people were using over 6,000 watts of power, equivalent to over 100 lightbulbs.[7]

CONSUMER CAPITALISM

The Industrial Revolution not only provided new production techniques, it also provided more efficient ways of killing people.[34]

The First World War (1914–18) was one of the deadliest in history, with more than 17 million deaths and over 20 million wounded.[35]

It was supposed to be the 'war to end all wars'.[36]

Yet, just 21 years later, Europe was at the centre of the Second World War.[35]

This war killed between 50 and 80 million people.[35]

At the end of the Second World War, the major Allies (USA, UK, Soviet Union and China) met to build the New World Order.[35]

In 1944, at the Dumbarton Oaks Conference in Washington DC, they laid the foundation for the 1945 Charter of the United Nations and the UN Security Council.[35]

In 1944 all 44 Allied countries met in Bretton Woods, New Hampshire, and agreed to control exchange rates by fixing them to the dollar, itself fixed to the price of gold.[35]

The Bretton Woods Conference created the International Monetary Fund to loan money to countries, helping maintain exchange rates and the global economy.[35]

A new global financial architecture was born.[35]

The United States became the world's most powerful country, using soft power to push cultural values, including mass consumerism with its ever-increasing environmental impacts.[37]

The General Agreement on Tariffs and Trade (GATT) came into effect in 1948 to enable the liberalization of global trade.[38]

The Great Acceleration had begun.[39]

IN 1950 THE GLOBAL POPULATION WAS 2.5 BILLION

World trade increased by over 6% per year from 1948 until 2005, when GATT was replaced by the World Trade Organization.[38]

The Cold War between the West and the Soviet bloc produced an economic arms race to industrialize fastest and furthest.[40]

This led to the largest investments in science the world had ever seen as each side tried to gain a competitive advantage.[40]

In 1986 we reached peak 'mutually assured destruction', with 69,368 nuclear weapons deployed and ready to use.[41]

Scientific insights now threatened to end human civilization.

The development of new medicines, improved living conditions and the Green Revolution in agriculture reduced infant mortality and produced ever more food.[42]

In 1950 the global population was 2.5 billion.[43]

In 2020 the global population was 7.8 billion people.[43]

A rise of over 5 billion people in 70 years.[43]

Energy use rose rapidly during the Great Acceleration to fuel this ever-growing population.[42]

IN 2020 THE GLOBAL POPULATION WAS 7.8 BILLION PEOPLE

An average American now uses over 10,000 watts of power for their cars, homes, offices and the rest of their lives, equivalent to running about 160 old-fashioned lightbulbs – compared to just 6 lightbulbs equivalents used by our hunter-gatherer ancestors.[7]

Today, humanity directly uses about 17 trillion watts of power – 280 billion lightbulbs.[44]

This is equivalent to the energy captured from photosynthesis in half the world's rainforests.[7]

Until 1980, governments continually intervened in the international economic system, investing when recessions occurred and regulating during boom times.[45]

Since the 1980s and the fall of communism, neoliberalism has been the dominant economic theory in Western government policies.[45]

Less government intervention, less regulation, less control, more individual freedom, more global trade, less support for the worse off in society.[45]

Until 2008, when governments had to bail out the global economic system.[45]

TODAY, HUMANITY DIRECTLY USES ABOUT 17 TRILLION WATTS PER DAY – 280 BILLION LIGHTBULBS

And 2020, when governments shut down the economy to protect people from Covid-19.[46]

Atmospheric carbon dioxide levels rose from 310 ppm during the Second World War to over 412 ppm in 2020.[33]

A rise of 1.3 ppm per year, 4 times faster than in the previous 100 years and 400 times faster than in the last 5,000 years.[33]

CHAPTER 3:
STATE OF OUR WORLD

HUMANITY IS THE NEW GEOLOGICAL SUPERPOWER.

WE NOW CONTROL THE ENVIRONMENT AND THE EVOLUTION OF LIFE ON EARTH.

For the first time in our planet's 4.5-billion-year history a single species, humans, is dictating its future.[1]

We are having as much effect on the Earth's environment as a giant meteorite, mega-volcanoes or the movement of the continental plates.[1]

We are the new geological superpower and have driven Earth into a new geological epoch: the Anthropocene.[1]

We have made enough concrete to cover the whole surface of the Earth in a layer 2 mm thick.[2]

We move more soil, rock and sediment each year than all the natural processes combined.[3]

We have created over 170,000 synthetic mineral-like substances, such as all plastics, concrete, steel, ceramics and many artificial drugs. (There are approximately 5,000 'natural' minerals.)[4]

There are 1.4 billion motor vehicles, 2 billion personal computers and more mobile phones than people on Earth.[5]

**WE MOVE MORE
SOIL, ROCK AND
SEDIMENT EACH YEAR
THAN ALL NATURAL
PROCESSES COMBINED**

THERE ARE MORE LEGO MINI-PEOPLE IN THE WORLD THAN REAL PEOPLE

We make over 300 million tonnes of plastic per year,[6] equivalent in weight to 1 billion African elephants or every person on Earth.

Waste plastic can be found in every ocean on our planet. A plastic bag has even been found in the Mariana Trench (10,984 m/7 miles deep).[7]

There are more Lego mini-people in the world than real people.[8]

Factories and farms fix more nitrogen in the atmosphere than all naturally occurring processes combined – the last time the nitrogen biogeochemical cycle was this seriously disrupted was over 2 billion years ago when oxygen first started to rise in the atmosphere.[9]

You are one of 7.8 billion humans in the world.[10]

Since the beginning of civilization we have cut down 3 trillion trees, more than half the trees on Earth.[11]

The current weight of all land mammals in the world is made up of 30% humans, 67% livestock and 3% wild animals.

10,000 years ago wild animals made up 99.95% of the weight.[12]

IN THE LAST 400 YEARS THERE HAVE BEEN 789 CONFIRMED SPECIES EXTINCTIONS

We extract 80 million tonnes of fish from the sea every year. Another 80 million tonnes are farmed.[13]

Farmland annually produces 4.9 billion head of livestock and over 4.9 billion tonnes of the top 5 crops: sugar cane, maize, rice, wheat and potatoes.[14]

Since AD 1500 there have been 789 documented species extinctions, including 79 mammals, 129 birds, 21 reptiles, 34 amphibians, 81 fish, 359 invertebrates and 86 plants.[15]

Our industry, farming and land-use changes have increased carbon dioxide levels in our atmosphere by over 47% and methane by over 250% since the beginning of the Industrial Revolution in the mid 18th century.[16, 17]

We have added 2.2 trillion tonnes of carbon dioxide to the atmosphere since the beginning of the Industrial Revolution, equivalent in weight to 20 Great Walls of China.[18]

25% of this extra 'anthropogenic' carbon dioxide has come from the USA.[19]

22% of this extra carbon dioxide has come from the EU.[19]

Less than 5% has come from Africa.[19]

OUR ADDITIONAL GREENHOUSE GASES HAVE INCREASED THE TEMPERATURE OF THE EARTH BY OVER 1°C

Carbon dioxide levels in the atmosphere are now higher than they have ever been in the last 3 million years.[20]

Greenhouse gases are a natural part of the Earth's climate system.[21]

Greenhouse gases absorb and re-emit some of the heat radiation given off by the Earth's surface, warming the lower atmosphere.[21]

The most common and powerful greenhouse gases found in our atmosphere, in order of impact, are: water vapour, carbon dioxide, methane, nitrous oxides and chlorofluorocarbons (CFCs).[21]

Without greenhouse gases in the atmosphere the Earth's average surface temperature would be approximately -20°C.[21]

But our additional greenhouse gases have increased the temperature of the Earth by over 1°C and caused sea levels to rise by over 20 cm since the beginning of the 20th century.[21]

Increasing carbon dioxide levels in the atmosphere is causing the oceans to acidify, disrupting life in the oceans.[22]

Our oceans are warming and have already lost over 2% of their dissolved oxygen, which is essential to marine life.[23]

The science of climate change is over 180 years old.[24]

Evidence for climate change is unequivocal.[25]

Natural events such as sunspots and volcanic eruptions have influenced the pattern of temperature changes over the past 150 years, but the overall warming trend is due to human-derived greenhouse gases.[25]

Significant changes in the Earth's climate system have been observed:

warming of the land all over the world,[26]
warming of the oceans,[27]
sea levels rising around the world,[28]
reduced snowfall in the Northern Hemisphere,[29]
retreating sea ice in the Arctic,[30]
retreating glaciers on all continents,[31]
shrinking of the Greenland ice sheet,[32]
melting of the Antarctic ice sheet,[33]
melting of permafrost,[34]
earlier occurrence of plant growth in spring,[35]
changes in bird migration patterns,[36]
shifts in the geographic ranges of some plants and animals.[37]

All of these are consistent with a warming global climate.[25]

THE SCIENCE OF CLIMATE CHANGE IS OVER 180 YEARS OLD

EVIDENCE FOR CLIMATE CHANGE IS UNEQUIVOCAL

THE SCIENCE OF CLIMATE CHANGE IS OVER 180 YEARS OLD

Weather patterns have shifted and extreme weather events have increased all around the world, including:[25]

> super-storms,[38]
> mega-floods,[39, 40]
> severe droughts,[41]
> unprecedented heatwaves,[42]
> uncontrollable wildfires.[43]

Between 1880 and 2020, the 19 warmest years on record have all occurred within the past 20 years.[44]

> 2016 is the warmest year ever recorded.[44]

> 2020 is the second warmest year on record.[44]

We are not all equally liable for the mess we find ourselves in.

The richest 10% of the world's population emit 50% of carbon pollution into the atmosphere.[45]

The richest 50% of the world's population emit 90% of carbon pollution into the atmosphere.[45]

The poorest 3.9 billion people have contributed just 10% of the carbon pollution in our atmosphere.[45]

THE RICHEST 50% OF THE WORLD'S POPULATION EMIT 90% OF CARBON POLLUTION INTO THE ATMOSPHERE

THE POOREST 3.9 BILLION PEOPLE HAVE CONTRIBUTED JUST 10% OF THE CARBON POLLUTION IN OUR ATMOSPHERE

780 million people live on less than $1.90 a day.[46]

Half of these people live in just five countries: India, Nigeria, Democratic Republic of Congo, Ethiopia and Bangladesh.[46]

4.5 billion people live on less than $10 a day.[46]

Poor people in developing countries can spend up to 80% of their income on food.[47]

Americans spend less than 10% of their income on food.[48]

75% of the world's food is generated from only 12 plants and 5 animal species:[49]

Plants: cassava, corn, maize, plantains, potatoes, rice, sorghum, soybeans, sugar cane, sweet potatoes, wheat and yams.[49]

Animals: Cows, pigs, chickens, sheep and goats.[49]

We produce enough food to feed 11 billion people.[50]

There are 7.8 billion people in the world today.[10]

7 million children die each year of hunger and preventable diseases.[51]

825 million people do not have access to enough food – this has increased from 800 million 5 years ago.[47]

Americans waste about 141 trillion calories worth of food every day. That adds up to approximately $165 billion per year: 4 times the amount of food Africa *imports* each year.[52]

More than 9 out of 10 farms in the world are small family-run farms. They provide 80% of the world's food supply.[53]

Agriculture provides jobs for around 30% of the world's population, making it the single largest employer in the world.[54]

An eighth of the world's population, nearly 1 billion people, live without electricity.[55]

This is the state of our planet.

This is the century in which we must set things right.

CHAPTER 4:
TAXONOMY OF DENIAL

THE GODS THEMSELVES STRUGGLE VAINLY WITH THE RIGHTEOUSNESS OF SELF-INTEREST.

Know thine enemy.

Counter their arguments.

Make them your ally or make them irrelevant.

The fossil fuel industry, political lobbyists, media moguls and individuals have spent the last 30 years sowing doubt about the reality of climate change – where none such doubt exists.

The world's 5 largest publicly owned oil and gas companies spend approximately $200 million per year on lobbying to control, delay or block binding climate-motivated policy.[1]

Ignoring the environmental crises facing the planet and our species does not make it go away. It makes it worse and ultimately more costly to fix.[2]

Politics is no excuse for a lack of action. Solutions to climate change can be found across the whole political spectrum.

Our planet does not care what colour party badge you wear or who you vote for, only that there is action.

Beware of the four subtle flavours of *denial* . . .

SCIENCE DENIAL

When it snows, the deniers declare there is no climate change.

Climate change will not stop winter weather – though it may make winters warmer and wetter with more frequent storms.[3]

Deniers will tell you that climate change is just part of the natural cycle and that climate is always changing.

This is deeply misleading.

Over the past 2,000 years, only in the last 150 years has the climate all around the world changed at the same time and in the same direction with warming of over 98% of the surface of the planet.[4]

This is not natural.

Deniers say that extreme weather events are just weather.

Attribution science has studied 113 extreme weather events in the last 5 years of which 70% were more likely due to climate change, 26% were less likely and 4% showed no variation due to climate change.[5]

Deniers say that climate change is due to sunspots or galactic cosmic rays (GCRs).

There is no scientific evidence for this statement.

Sunspot activity has not caused an upward trend in the amount of the Sun's energy hitting the Earth.[6]

GCRs have been shown to have very little effect, if any, on climate.[7]

Deniers will tell you that CO_2 is a small part of the atmosphere so it can't have a large heating effect.

This is both false and nonsensical.

For over 150 years scientists have measured the ability of atmospheric carbon dioxide to retain heat both in the laboratory and in the atmosphere.[8]

Just because something is present in a small quantity doesn't mean it will have a small effect. 0.1 grams of cyanide, approximately 0.0001% of your body weight, will kill you.[9]

Carbon dioxide currently makes up 0.04% of the atmosphere and is a strong greenhouse gas. Nitrogen makes up 78% of the atmosphere and is highly unreactive. Effect, not size, is what matters.[10]

Deniers will tell you that scientists manipulate all data sets to show a warming trend.

This is not true and would be highly impractical and unlikely, requiring thousands of scientists in more than 100 countries to conspire.[11]

Deniers say that climate models are unreliable and too sensitive to carbon dioxide.

There is no evidence for this.

No single model should ever be considered correct – they represent a very complex global climate system. But collectively since the 1970s they have reliably predicted the steep 1°C warming of the last hundred years.[12]

ECONOMIC DENIAL

Deniers will tell you that climate change is too expensive to fix.

This is false.

Last year the world generated $88,000,000,000,000, reflecting an average growth of 3.5% year on year.[13]

Economists suggest we could fix climate change now by spending 1% of world GDP.[2]

This cost could be less if we count the cost savings due to improved human health and expansion of the global green economy.[14]

If we don't act now, by 2050 it will cost over 20% of world GDP.[2]

Climate change is too expensive *not* to fix.[2]

Currently the fossil fuel industry receives $5.2 trillion in subsidies.[15]

The largest subsidizers are: China ($1.4 trillion), United States ($649 billion), Russia ($551 billion), European Union ($289 billion) and India ($209 billion).[15]

This amounts to 6% of world GDP.[15]

Imagine the positive things you could do with this money.

**BY 2050
CLIMATE CHANGE
COULD COST
OVER 20% OF
WORLD GDP**

HUMANITARIAN DENIAL

You will hear deniers argue that climate change is good for us.

This is false.

Deniers say longer warmer summers in the temperate zone will make farming more productive.

These small gains are offset by the impact of extreme weather, such as prolonged droughts, mega-floods and the increased frequency of heatwaves.[16–19]

The Moscow heatwave devastated the Russian wheat harvest in 2010 and caused global food prices to increase significantly.[20, 21]

Deniers say atmospheric carbon dioxide is a plant fertilizer, so more is better.

This is a small but measurable effect.[22]

The land biosphere has already been absorbing about a quarter of our carbon dioxide pollution every year.[23]

Another quarter of our carbon pollution is absorbed by the oceans.[24]

Unfortunately, an area the size of the UK is deforested every year, which completely nullifies this minor fertilization effect.[25]

Deniers will tell you that more people die of the cold than heat – so warmer winters due to climate change will be a good thing.

This is false.

Vulnerable people die of the cold because of poor housing and not being able to afford heating bills. Society, not climate, kills them.[26]

In the USA over the last 30 years more than 4 times the number of people died from heat than cold.[27]

Heatwaves cause cumulative physiological stress on the human body, which exacerbates respiratory and cardio-vascular diseases, diabetes mellitus and renal disease.[28]

There are *no* positive effects of anthropogenic climate change.

POLITICAL DENIAL

Deniers will tell you that we cannot take action because other countries are not taking action.

Some countries have a greater greenhouse gas legacy and have done more to cause current climate change.[29]

Western countries have produced over half the extra carbon dioxide in the atmosphere, while India has produced just 3%.[29]

The historic legacy of greenhouse gas pollution means developed countries have a responsibility to lead the way.[30]

All countries need to act.

Every country must go net carbon zero by 2050 if we are to avoid a temperature increase of over 2°C. If we do it fast enough, we could even keep the global temperature rise to just 1.5°C.[31]

Is it a burden to make your country a better place to live?

Switching to renewable energy and electric vehicles reduces air pollution, improving people's overall health.[32]

Developing a green economy provides economic benefits and creates jobs .[33]

The global green economy is currently over $10 trillion per year.[33]

In the US the green economy employs 9.5 million people – nearly 10 times the amount in the fossil fuel industry.[33]

Improving the environment and reforestation provides protection from extreme weather events and can improve food and water security.[34]

Dealing with climate change makes the world a safer, healthier and fairer place to live, for everyone.

CHAPTER 5:
POTENTIAL FUTURES – NIGHTMARE OR ECOTOPIA?

HELL HAS NO FURY LIKE NATURE ABUSED BY ITS GUARDIANS.

NIGHTMARE – YEAR 2100

Our children feel contempt at our failure to look after our planet.

This section shows you the terrifying vision of our planet in 80 years' time if we do nothing about the climate and environmental crisis facing us now.

HEAT

Global temperatures have risen by over 4°C at the end of the 21st century.[1]

In many countries summer temperatures persistently stay above 40°C.[2]

Heatwaves with temperatures as high as 50°C have become common.[2]

GLOBAL TEMPERATURES HAVE RISEN BY OVER 4°C AT THE END OF THE 21ST CENTURY

Working outside has become physiologically impossible for many days each year due to the increased heat and humidity.[3]

Every summer wildfires rage across: Australia, Argentina, Brazil, California, Canada, Central America, Indonesia, India, Mongolia and North China, Southern Africa, Russia, Sub-Saharan Africa, Texas and all around the Mediterranean.[4]

Wildfires have created major air pollution events and human health crises.[5]

Ocean temperatures have risen dramatically and repeated bleaching events means the Great Barrier Reef has been officially declared dead.[6, 7]

DROUGHT

Large areas of the Earth are experiencing frequent prolonged droughts.[8]

The deserts of the world have expanded, displacing many millions of people.[9]

3.5 billion people live in areas where water demand is greater than the water availability.[10]

3.5 BILLION PEOPLE LIVE IN AREAS WHERE WATER DEMAND IS GREATER THAN THE WATER AVAILABILITY

An extra 1.5 billion people are water-stressed compared to 2020.[10]

Large areas of the world become unsuitable for agriculture due to lack of rainfall.[11]

Air pollution has increased due to the huge increase in dust from the expanded arid areas.[9]

ICE

The Arctic is free of sea ice every summer.[12]

Arctic temperatures have risen by over 8°C due to the lack of sea ice.[1]

The Greenland and Western Antarctic ice sheets have started to melt, releasing a huge amount of freshwater into the oceans.[13]

Many mountain glaciers have completely melted.[14]

Skiing now takes place on giant artificial slopes.[15]

The highest mountain in Africa, Kilimanjaro in Tanzania, has no ice.[16]

THE ARCTIC IS
FREE OF SEA ICE
EVERY SUMMER

Ernest Hemingway's short story 'The Snows of Kilimanjaro' exists only as a memory of lost times and the climate of East Africa has irreversibly changed.[16]

Most of the Himalayan plateau ice has melted, reducing the flows of the Indus, Ganges, Brahmaputra and Yamuna rivers that used to provide plentiful and essential water for over 600 million people.[17]

SEA LEVEL

Thermal expansion and melting ice sheets have caused sea levels to rise by over 1 m.[18]

Many major cities are already flooded and are uninhabitable including:

Asia: Dhaka (20.3 million people today), Shanghai (17.5 million), Hong Kong (8.4 million) and Osaka (5.2 million).

North America: Miami (2.7 million).

South America: Rio de Janeiro (1.8 million).

Africa: Alexandria (3.0 million).

Europe: The Hague (2.5 million).[19]

SEA LEVELS HAVE RISEN BY OVER 1 M

These cities are just the start – over 40% of the world's population live less than 60 miles from a coast and will be affected by extreme sea level rise.[20]

The Maldives, the Marshall Islands, Tuvalu and many other small island nations have been abandoned.[20]

Many coastal and river areas are regularly flooded, including: Nile Delta, Rhine Valley, Nigeria, Thailand, Mississippi, Mekong, Mahanadi, Godavari and Krishna.[20]

Over 20% of Bangladesh is underwater.[21]

London is now protected by a new 13km-long barrier stretching from Kent to Essex.[22]

Many cities are not so lucky.

And this is just the start – the melting of the Greenland and Western Antarctic ice sheets is now irreversible and in the next two centuries sea level could rise by over 15 m.[23]

STORMS

Winter storms have become more intense, causing widespread damage and flooding.[24]

OVER 20% OF BANGLADESH IS UNDERWATER

Tropical cyclones (hurricanes, typhoons and cyclones) have become stronger and more intense, affecting tens of millions of people every year.[25]

Mega-cyclones like Typhoon Haiyan (2013) have become commonplace, with sustained wind speeds of over 200 mph.[26]

South-East Asian monsoons have become more intense and variable, bringing either too much or too little water to each region, affecting the lives of over 3 billion people.[27]

FLOODS

Rainfall has become more seasonal and falls in more intense bursts.[1]

Flash flooding is a major issue in most urban areas.[28]

Most coastal areas are prone to frequent flooding due to sea level rise and stronger storms.[18]

Many coastal areas and floodplains have been abandoned as it is no longer possible to protect people from floods.[18]

OCEAN ACIDIFICATION

The acidity of the ocean has dropped to 7.8 pH units, an increase of 125% in the number of hydrogen ions (H+) in seawater, which causes acidity.[29]

The ocean food chain has collapsed in some regions as marine organisms struggle to make calcium carbonate shells in the more acidic waters.[30]

FOOD

Food security, the reliable access to a sufficient quantity of affordable, nutritious food, has become a major issue.[31]

Extreme heat and humidity in the tropics and subtropics has increased the number of days that it is impossible to work outside tenfold – limiting food production.[3]

In the temperate zone, squeezed between the hot tropics and the cold polar regions, frequent extreme weather events have made food production highly variable.[31]

Half of the land devoted to agriculture in the past is now un-usable, and the rest is highly volatile from season to season.[31]

Crop yields are the lowest they have been since the middle of the 20th century.[32]

Fish stocks have collapsed from overfishing and ocean acidification.[32]

Half the world population is still too poor to buy basic food when food prices spike due to unpredictable yields.[31]

Famines have become commonplace.[33]

HUMAN HEALTH

Food and water insecurity threatens the health and wellbeing of billions of people.[34]

Despite advances in medical sciences, deaths from tuberculosis, malaria, cholera, diarrhoea and respiratory illnesses have risen to their highest levels.[35]

Extreme weather events (heatwaves, droughts, storms and floods) are causing large loss of life, creating millions of homeless people and increasing food and water insecurity.[35]

New pandemics have started to occur, spreading faster because of widespread poverty and vulnerability.[36]

MIGRATION

People are moving from the tropics and subtropics into the temperate zone.[37]

People are migrating from Central America through Mexico towards the United States.[38]

People are migrating from Africa and the Middle East towards Europe.[39]

In Bangladesh people are migrating towards India.[40]

These mass movements of people are repeated all around the world.[37]

Refugee camps, internment camps, violence, civil unrest, civil wars and frequent outbreaks of disease dominate the daily news broadcasts.[41]

*   *   *

This could be our brave new world – if we do nothing.

ECOTOPIA – YEAR 2100

Our children rejoice and celebrate us because we saved our planet.

This section shows you what our planet could be like by the end of the 21st century if we do everything we can to deal with the climate and environmental crisis facing us now.

GLOBAL COOL

Global temperatures rose by just 1.5°C at the end of the century.[42]

Fossil fuels have been replaced by clean safe renewable energy.[43]

Over a trillion trees have been planted.[44]

The air is cleaner than it's been since before the Industrial Revolution.[42]

Cities have been restructured to provide excellent electric public transport and many new vibrant green open spaces.[45]

Buildings all now have a photoelectric/photovoltaic skin to generate solar energy.[46]

Many buildings now have green roofs to cool the city and make it a more pleasant place to live.[46]

High-speed electric trains, going over 300 mph, link many of the major cities of the world.[47]

Intercontinental flights still connect the world using very large efficient planes fuelled with synthetic kerosene.[48]

New technology makes virtual meetings feel almost real and these have become the new norm, greatly reducing business travel.[49]

Pandemics of the early 21st century helped shift global diets to be much more vegetable-based, helping to improve everyone's health.[50]

Most countries in the world rolled out guaranteed basic payments to all adults, lifting billions out of poverty.[51]

Farming efficiency has been greatly improved and shifted away from meat production, creating more land to rewild and reforest.[31]

The 'half-Earth' idea has been used as a fundamental principle to guide international organizations so that:

Half of the Earth is used to support the 10 billion people on the planet. [52]

Half of the Earth is used to support the natural biosphere and the ecological services we rely on.[52]

Finally, they think the technological difficulties of making fusion energy work have been resolved, providing unlimited clean energy for the 22nd century.[53]

If this is the world you want, then just read on . . .

OVER A TRILLION TREES HAVE BEEN PLANTED

CHAPTER 6:
POWER OF THE
INDIVIDUAL

SOLVE CLIMATE CHANGE.

TOGETHER WE MUST.

Throughout history it is individuals that have changed the world.

A young girl standing outside the Swedish Parliament protesting about climate change.

A black woman refusing to give up a seat for a white man.

Individuals are the catalysts that allow the rest of us to demand change.

Here are 15 suggestions of how *you* can help change the world.

1. TALK ABOUT CLIMATE CHANGE.[1]

The first and most important thing you can do is talk about climate change to everyone.

The greatest challenge in the history of our species should not be a taboo subject.

INDIVIDUALS ARE THE CATALYSTS THAT ALLOW THE REST OF US TO DEMAND CHANGE

We need new solutions, new social structures and new economics to solve this crisis.

So talk about climate change. Share the ideas in this book – just one idea shared with one of your friends or relatives will start the conversation rolling.

2. SWITCH TO A MORE VEGETABLE-BASED DIET.[2]

You can become a *flexitarian* – eating a diet of mostly plant-based foods with meat and dairy products in moderation.[3]

A standard Western meat-based diet produces 7.2 $kgCO_{2e}$ per day (all GHG emissions converted to CO_2 equivalent (e) to allow comparisons).[4]

A vegetarian diet produces 3.8 $kgCO_{2e}$ per day.[4]

A vegan diet produces 2.9 $kgCO_{2e}$ per day.[4]

Meat production, particularly beef, is a major cause of tropical deforestation. Cutting your meat consumption saves the environment and reduces your carbon emissions.[5]

A MORE VEGETABLE-BASED DIET IS GOOD FOR YOU AND YOUR FAMILY'S HEALTH

A MEAT DIET PRODUCES 7.2 KgCO$_{2E}$ PER DAY

A VEGETARIAN DIET PRODUCES 3.8 KgCO$_{2E}$ PER DAY

A VEGAN DIET PRODUCES 2.9 KgCO$_{2E}$ PER DAY

A more vegetable-based diet is also good for you and your family's health.[5]

Meat, especially highly processed meat, has been linked to: high blood pressure, heart disease, chronic obstructive pulmonary disease, bowel and stomach cancer.[5]

Use locally sourced and seasonally appropriate food to reduce carbon air miles and support your local economy.[5]

If you're not already, aim to become vegetarian.[6]

Try dairy-free alternatives.[7]

3. SWITCH TO A RENEWABLE ENERGY SUPPLIER.[8]

A simple change that may not even cost you anything extra.[9]

If we all switched, then energy companies would have to generate more renewable energy to meet demand.

Persuade your work, place of worship, local authority, school and sports club all to switch to renewable energy.[10]

4. MAKE YOUR HOME ENERGY EFFICIENT.[11]

Save energy by making your home as heat efficient as possible.

Ensure your home is insulated (roof and walls) to the highest standard.[12]

Ensure your windows and doors are fitted properly to avoid draughts.[12]

Turn the thermostat down a degree, wash at a lower temperature, reduce the use of home appliances, install a smart meter, replace halogen bulbs with more energy-efficient LED bulbs.[11]

Energy efficiency will save you money.[13]

If your heating system at home needs replacing, make sure at the very least it is replaced with the most efficient gas boiler possible – but better still invest in ground and air heat exchangers that can produce either heating or cooling (which you will need as the world heats up).[12]

5. USE CARS LESS.[14]

Increase your walking, cycling and use of public transport.[15]

This will improve your fitness and thus your health.[16]

If you need a car occasionally then hire an electric or hybrid one.[17]

If you really need a car, choose the smallest and most efficient one possible or, if you can afford it, buy an electric or hybrid car.[18]

6. STOP FLYING.[19]

Choose an alternative form of travel such as the train.[20]

If you must fly for work, then select only essential trips and ensure you offset the carbon emissions using reputable specialist firms.[21]

Alternatively use the 'offset' money to reduce your emissions at home or work.

Many organizations have now decided to offset all unavoidable emissions at 10 times the estimated emissions and to monitor closely their chosen offset schemes. This way they can be assured their emissions have been removed.[21]

7. DIVEST YOUR PENSION FROM FOSSIL FUELS.[22]

Lobby your pension fund to divest from all fossil fuel investments.

Or move your pension fund if they do not divest.

Your pension money will be safer and may even earn more.

8. DIVEST YOUR INVESTMENTS FROM FOSSIL FUELS.[23]

The fossil fuel industry will be greatly affected by future climate change legislation and will cease to be profitable.

Fossil fuel companies will suffer from what are called 'stranded assets'.[23]

The company's nominal worth is based on how much fossil fuels they have in their oil and gas fields and their coal mines. These assets will be worthless when fossil fuel mining and drilling are banned due to climate change.[24]

Fossil fuels are a poor long-term investment.[24]

Green companies are already returning twice as much profit as fossil fuel companies.[25]

9. REFUSE/REJECT EXCESSIVE CONSUMPTION.[26]

You do not need all that stuff.

Reject the idea that consumption is good for you and that more stuff makes you happier.

Think carefully about what you need and how you could achieve a sustainable low-carbon lifestyle.

Competing to have the latest gadget, car or clothes just leads to social stress.

Not worrying about what everyone else is buying will reduce your social stress and make you healthier and happier.[27]

Friends, family, work and community can make you happy.

**AS A CONSUMER
YOU HAVE HUGE
POWER, EXERCISE
THAT POWER THROUGH
YOUR CHOICES**

DO YOU NEED
FAST FASHION?

10. REDUCE WHAT YOU USE.[28]

Make sure that you encourage everyone you know to reduce what they use.

As a consumer you have huge power, exercise that power through your choices.[29]

Do you need so much packaging on your goods?

Do you need fast fashion?

Do you need plastic bottles or throwaway coffee cups?

Do you need a big SUV car or a truck?

Plan what you eat to reduce food waste in your household.

11. REUSE AS MUCH AS YOU CAN.[28]

Buy clothes that are stylish but will last, so you reuse them again and again.

Have a water bottle and a reusable coffee cup.

Engage with your local community reuse network – someone local may want to reuse the stuff you no longer need or want.

Fix things when they break.

Replace your phone battery and/or screen instead of buying a whole new phone.

12. RECYCLE AS MUCH AS YOU CAN.[30]

The circular economy is essential if we are to have a sustainable economy so we must recycle as many materials as we can to reduce the high-energy production and environmentally destructive extraction of raw materials.[31]

Separate recyclable materials and compost from non-recyclable materials.[13]

If possible, compost your organic material and food waste to use on your garden or vegetable patch.[13]

Try to reduce the amount of non-recyclable materials you use – but they can be burned to recover energy through high energy combustion, so they are still part of the circular economy.[31]

If your municipal authority or local council does not take recycling or energy recovery seriously, lobby, protest and make them serve the community properly.[30]

13. USE YOUR CONSUMER CHOICE.[32]

Consider the carbon footprint of products (air miles and environmental impacts, e.g., deforestation).

Buy locally, eat seasonally or try growing your own.

Support environmentally responsible and sustainable companies.

14. PROTEST.[33]

People power is real.

The School Climate Strikes and the Extinction Rebellion protests have brought together diverse groups of people across the world all wanting governments to start taking the protecting of our planet seriously.[34, 35]

Protests are having an impact and change is starting to happen.[36]

PEOPLE POWER
IS REAL

DEMAND CHANGE

15. VOTE.[37]

In democratic countries you can vote for new governments.

Demand your politicians tell you where they stand on dealing with climate change.

Demand they tell you where they are getting their campaign money.

Demand change.

Demand action.

Use your vote wisely.

DEMAND ACTION

CHAPTER 7:
CORPORATE
POSITIVE POWER

WE DO NOT HAVE TIME FOR THE REVOLUTION:

WE MUST USE CURRENT SYSTEMS TO START TO SAVE OUR PLANET.

The world's top 100 companies generate more than $15 trillion in revenue per year.[1]

More than half of the world's 100 most valuable companies are located in either the US (35 companies) or China (23 companies).[1]

Businesses control our lives: what we eat, what we buy, what we watch and even who we vote for.[2]

Corporations have immense power and we must harness this to change the world for the better.[3]

They are one of our major weapons of change.[4]

Below are suggestions of what you can do within your corporation or organization to help them help us save our planet.

1. AMBITION AND VISION

Short-termism in the corporate world is so last century.[2]

Companies need to set the agenda as they are more agile and fast-moving than governments.[3]

Being sustainable and actively caring about the environment is good for business.[4]

Corporations that actively manage and plan for climate change secure a 67% higher return than companies who refuse to disclose their carbon emissions.[5]

Microsoft has set the agenda for the technology sector with the ambition to go carbon negative by 2030.[6]

By 2050 they want to have removed all the carbon pollution from the atmosphere that they and their supply chain have emitted since the founding of the company in 1975.[6]

Sky has set the agenda for the media sector, pledging that they and their supply chain will be zero carbon by 2030.[7]

BP has also declared that it will be carbon neutral by 2050 by eliminating or offsetting over 415 million tons of carbon emissions.[8]

2. OPEN AND TRANSPARENT BUSINESS

Going carbon neutral or carbon negative is a journey that needs to be transparent so that employees, clients, customers and regulators can appreciate what is being done.[9]

Encourage your company to measure its carbon footprint.

Help your company plan how it will reduce its carbon footprint.

Encourage them to publish their carbon footprint and how they plan to reduce it.

Help your company to implement their plan.

Encourage the company to publish its successes and failures.

Actively promote your company's successes on social media.

Get them to repeat this cycle until they are carbon negative.[10]

You can apply the same process to other greenhouse gases, water usage and raw material consumption – remember it is *your* company.[11]

3. SET EMISSION REDUCTION TARGETS

Setting targets is essential to reducing carbon emissions and environmental damage.[12]

Ensure that your company's targets comply with European and international level regulations on reporting greenhouse gas emissions.[13]

Get your company to set an ambitious long-term emission reduction target that has been approved by the Science Based Targets initiative (SBTi).[13]

Ensure your company's reporting conforms to the Task Force on Climate-related Financial Disclosures (TCFD).[14]

Ensure your company publishes an annual environmental audit to the most stringent international standards.[14]

Suggest your company sets an internal price on carbon to motivate behavioural change.[14]

Get your company to aim to be a 'carbon neutral' company.[10]

Continually increase your targets to become a 'carbon negative' company.[10]

Encourage your company to calculate the total carbon footprint since it was founded and ultimately aim to remove all that carbon pollution.[10]

4. FOCUS ON ENERGY

Energy consumption is a significant source of greenhouse gas emissions for most companies.[15]

Energy is central to the manufacturing of products and servicing of offices, factories, warehouses and/or data centres.[15]

Measure and monitor your company's energy use.[16]

Use only the highest-rated green buildings available.[17]

Move to renewable energy for electricity through energy suppliers and power purchase agreements.[18]

Generate onsite renewable energy where practical.[18]

Switch to renewable energy instruments such as Guarantee of Origin (GO) and International Renewable Energy Certificates (I-REC & REC) for electricity.[18]

Switch to Green Gas Certificates – every unit of green gas injected into the grid displaces a unit of conventional gas.[19]

Switch your company's car/van fleet to electric.[20]

Switch your truck fleet to biodiesel in the short term and to electric in the medium term.[21]

5. APPLY THE CIRCULAR ECONOMY

The classic economic model 'take, make, dispose' relies on large quantities of cheap, easily accessible materials and energy.[22]

We are reaching the physical limits of this model.[22]

The circular economy minimizes the amount of resources that are extracted and maximizes the value of products and materials throughout their lifecycle.[23]

Applying a circular economy could unlock up to €1.8 trillion of value for Europe's economy.[24]

Plan and make products that have longevity, upgradability and built-in recyclability.[25]

Design out waste and pollution in your company and products.[25]

Make products and materials that will be used for a long time.[25]

6. EMPLOYEE POWER

CEOs are taking sustainability seriously to ensure recruitment and retainment of the very best employees.[26]

Brilliant young people want to work for a company that has a strong moral and ethical code of conduct.[27]

Making companies environmentally sustainable requires profound changes in the company's own behaviour, harnessing the creativity of the employees by building a more collaborative mode of operation.[28]

You need to establish a sound Environment Management System (e.g. ISO 14001) to focus on the key aspects and impacts (energy use, business travel, waste, water etc).[29]

Employee power improves efficiency, innovation and environmental performance and can be grown through employee ambassadors, green champions, environmental specialists and partner organizations.[30]

Companies need to listen to their employees and, where practical, embrace and implement their ideas.[28]

Your company needs to encourage all the employees to embed environmental sustainability objectives into their daily work.[31]

Your company needs to support and encourage employee volunteering to promote the importance of environmental matters in the local community.[32]

7. LINK INTO THE SUPPLY AND VALUE CHAIN

In a globalized economy, companies rely on their supply chain to meet their business objectives.[33]

Your supply chain greenhouse gas emissions could be as much as 4 times that of your own company's (internal) emissions.[34]

Environmental innovation and sustainability need to be extended to *your* company's or organization's external value chain.[34]

This can be done by:

Creating a responsible purchasing policy, to promote the purchase of products and services with a low environmental impact.[34]

Engaging with your supply chain to ensure you work with companies that share your sustainable goals.[33]

Instigating key performance measures of your supply chain for sustainability.[35]

Encouraging technological developments in your supply chain to help them innovate and produce goods in a greener way.[36]

Using blockchain technology to track the environmental impact of the services and goods you purchase.[37]

Evaluating and auditing your supply chain annually against your key performance measures and your carbon emission target.[35]

8. CHANGE THE WHOLE CONVERSATION

If companies want to remain relevant and trusted in the 21st century, they need to change their relationship with the environment and society.[38]

Classic economics is dead.

The throwaway corporate culture is dead.

THE THROWAWAY CORPORATE CULTURE IS DEAD

The circular economy is essential if companies are to be part of saving our planet.[39]

Employees are the strength and the heart of all companies and organizations.

Sustainability must become a central tenet of all companies – this increases the longevity and profitability of the company.[5]

Make sure your company has a 5-year, 10-year, 15-year and 50-year plan.[40]

9. INFLUENCE GOVERNMENTS

Make your government aware that you take the future of our planet seriously.

Over 200 British firms have urged the UK government to align the post-Covid-19 economic recovery with net zero carbon goal, including Britvic, BP, BT, BNP Paribas, cement producer CEMEX, Coca-Cola, E.ON, Heathrow Airport, HSBC, IKEA, Lloyds Banking Group, Mitsubishi, National Grid, PwC, Severn Trent, Sky, Unilever, Shell and Siemens.[41]

MAKE YOUR GOVERNMENT AWARE THAT YOUR COMPANY TAKES THE FUTURE OF OUR PLANET SERIOUSLY

Politicians think about the short term; companies now think about the long term, so ensure new environmental policies will create long-term positive change.[42]

Companies rely on the infrastructure provided by governments. So lobby for it to be as sustainable, low carbon and high quality as possible.[43]

Companies need a highly skilled workforce to support the massively expanding global green economy. Lobby your government to have the very best education system possible.[44]

Governments set environmental regulations. Lobby your government to make sure these are sensible, strict, long-term and reward companies and organizations that comply fully.[42]

Once they have set strict environmental standards, make sure they stick to them, so your company can plan for the long term.[42]

Lobby your government to ensure independent (non-commercial) enforcement of all environmental regulations so that there is a level playing field to ensure fair competition between companies.[44]

Make sure your government is as forward-thinking and committed to saving our planet as your company.[42]

WE NEED DISRUPTORS, WE NEED NOVEL THINKERS, WE NEED NEW COMPANIES, TO DRIVE INNOVATION

WE NEED A
NEW WAVE OF SOCIAL
AND ENVIRONMENTAL
ENTREPRENEURS
TO HELP SAVE
OUR PLANET

10. CALLING ALL ENTREPRENEURS

The global green economy is growing at a phenomenal rate and needs new technology and new products.[45]

Rapid change brings about opportunity – never has there been a better and more important time to be an entrepreneur.[46]

> We need disruptors,
> we need novel thinkers,
> we need new companies,
> to drive innovation.

We need a new wave of social and environmental entrepreneurs to help save our planet.[46]

Is that you?

CHAPTER 8: GOVERNMENT SOLUTIONS

GOOD GOVERNANCE IS OUR GREATEST WEAPON AGAINST CLIMATE CHANGE.

Governments control the aspirations of civil society through the rule of law and the development of policy.

Incentives, subsidies, taxation and regulation all play a role in how governments can make our societies more sustainable and carbon neutral.[1]

Using win-win solutions they can make everyone's lives better.[2]

Governments drive innovation.[3]

All climate change solutions should be win-*win*.[2]

1. SUPPORT RENEWABLE ENERGY

Stop burning coal and natural gas.[4]

> *Reduces air pollution, saves lives and reduces healthcare costs.*

ALL CLIMATE CHANGE SOLUTIONS SHOULD BE WIN-*WIN*

Invest in solar, wind, hydro and tidal energy.[4]

> *Increases energy security by reducing dependency on imported fossil fuel.*

> *Creates new jobs in dynamic 21st-century industries.*

Ensure full energy recovery from non-recyclable waste by using high-energy incineration to generate heat and energy.[5]

> *Avoids sending waste to landfill, which can cause water pollution and create methane gas.*

Fund research and commercialization of fusion power.[4]

> *Long-term sustainable global energy demands in the 21st century could all be met using safe, clean fusion power.*

2. SUPPORT ELECTRIC CARS AND PUBLIC TRANSPORT

Invest in electric vehicle infrastructure.[6]

> *Reduces air pollution, saves lives and reduces healthcare costs.*

INVEST IN ELECTRIC VEHICLE INFRASTRUCTURE

INVEST IN PUBLIC TRANSPORT AND CYCLING

Creates new jobs in dynamic 21st-century industries.

Invest in and subsidize public transport and cycling.[7]

Encourages daily exercise, increasing health and wellbeing and reducing healthcare costs.

Reduces the average commuting time and increases wellbeing.

3. CUT FOSSIL FUEL INVESTMENT AND SUBSIDIES

Governments spend over $5 trillion per year subsidizing fossil fuel use.[8]

This money could be used to support healthcare, infrastructure projects, renewable energy and other essential government priorities.

Cut government investment in overseas fossil fuel projects and invest in sustainable renewable energy overseas instead.[9]

Supports decarbonization of other countries.

Stimulates international markets for home-grown renewable energy technology.

4. TAX FOSSIL FUEL USE

Tax the import and use of oil, coal and natural gas.[10]

Makes renewable energy and electric vehicles more competitive.

Increases government revenue for investment in low-carbon technology and infrastructure projects.

Tax internet companies internationally, based on their carbon emissions.[11]

Legitimizes taxation of international companies and incentivizes huge IT companies to reduce their carbon footprint.

Tax aviation fuel based on carbon emissions per mile.[10]

Drives efficiency and the development of new aviation technology.

Creates a financial incentive to use biofuels or carbon-neutral synthetic kerosene.

Reduces the stigma and guilt of flying – people will be able to choose more environmentally friendly flying.

5. BUILD NEW LOW-CARBON INFRASTRUCTURE

Build new electric high-speed 'bullet' train networks across the world.[7]

Replaces high-carbon flights with low-carbon trains.

Faster connectivity with reduced security risks.

Creates a huge number of jobs.

80% of all internal flights in the US could be replaced by high-speed train networks on the east and west coasts.

Design new building regulations so all new buildings are carbon neutral.[12]

Creates healthier homes, offices and factories.

Reduces the energy costs of homes, alleviating energy poverty.

Support low-carbon retrofitting of buildings.[13]

Creates healthier and cheaper buildings for people to live and work in.

Build new recycling facilities and generate energy from waste.[14]

Strengthens the circular economy, provides more low-carbon energy and prevents waste going to landfill.

6. REFOREST AND REWILD

Incentivize replanting of regionally appropriate mixed woodland or forest.[15-17]

Provides natural negative carbon dioxide emissions by storing carbon.

Increases and protects local biodiversity.

Stabilizes local soil.

Reduces the risk of extreme flash floods.

Stabilizes regional rainfall and weather patterns.

Increases productivity of adjacent agricultural land.

Aesthetically pleasing and can increase human wellbeing.

Incentivize rewilding of natural habitats such as wetlands, peatlands, grasslands, mangroves and areas of coastal seagrass.[18, 19]

Provides natural storage of carbon and reduces carbon emissions from the land.

Increases and protects local biodiversity.

Reduces the risk of extreme storm surges and flooding.

Reduces the rate of coastal erosion.

Increases productivity of adjacent agricultural land.

Aesthetically pleasing and can increase human wellbeing.

Use progressive land taxation to ensure the better use of land in protecting the environment.[20, 21]

> *Financially incentivizes reforestation and rewilding.*
>
> *Incentivizes low-carbon and sustainable farming.*
>
> *Incentivizes active flood management.*
>
> *Moves land ownership back into public ownership to ensure it is used productively for the public good.*

7. PROMOTE LOW-EMISSION FARMING AND DIET

Governments must provide a new vision of farming which protects the environment, increases food security, provides healthier diets and increases animal welfare.[22]

Mandate that food labelling must include the country of origin, carbon footprint, farming method and relative healthiness.[23]

> *Provides consumers with a real informed choice about the environmental impact and health implications of the food they buy.*

GOVERNMENTS MUST PROVIDE A NEW VISION OF FARMING WHICH PROTECTS THE ENVIRONMENT AND INCREASES FOOD SECURITY

Mandate that public procurement must provide only vegetable-based, highly nutritious and environmentally certified food.[24, 25]

> *Incentivizes large food companies to produce higher-quality vegetable-based food products.*

> *Increases revenue to environmentally positive farmers.*

> *Provides healthy diets for people who work for or are cared for by the state.*

Ban advertisement of unhealthy foods.[26] Tax unhealthy foods such as meat and sugar.[27, 28]

> *Makes unhealthy foods more expensive and less attractive.*

> *Helps change the food culture away from unhealthy food.*

Regulate and financially incentivize farming to restore soils, biodiversity and minimize water and air pollution.[17, 29-31]

> *Reduces greenhouse gas emissions from agriculture.*

Increases local biodiversity and helps maintain ecosystem services.

Can help to reduce the impacts of flash flooding and droughts.

Reduces the cost of water treatment and purification.

Encourage farming methods that work in harmony with nature such as agroecology, circular agriculture and organic farming.[32]

Reduces waste and greenhouse gas emissions from agriculture.

Increases local biodiversity and helps maintain ecosystem services.

Increases reforestation and rewilding within the agricultural sector.

Tax non-organic pesticides and synthetic fertilizers. Globally ban pesticides that are harmful to bees and other essential pollinators.[33]

Reduces waste and greenhouse gas emissions from agriculture.

Reduces the cost of water treatment and purification.

Increases local biodiversity and helps maintain pollinators, which are an essential ecosystem service.

8. SUPPORT THE EMISSIONS TRADING SCHEMES

Emissions trading or 'cap and trade' is a market-based environmental regulation that allocates permits, which companies have to buy if they want to emit greenhouse gases.[34]

These permits can be traded to ensure the companies that can reduce their emissions the fastest are financially incentivized to do so.[34]

The number of permits is cut each year and their price is increased to ensure a continual drop in greenhouse gas emissions and incentivize rapid innovation.[34]

Makes companies more efficient and environmentally friendly.

Saves companies and large organizations money.

Drives innovation and sharing of best practice between organizations.

9. ADAPTATION – PROTECT YOUR CITIZENS

Undertake and publish regular national climate change risk assessments.[35]

> *Allows companies, organizations, local government and individuals to assess the risks of climate change and adapt appropriately.*

Inform and educate the public on the enhanced risks due to climate change.[36]

> *Help citizens understand the risks of climate change and the actions taken by government to protect them.*

Support local infrastructure and social adaptations to climate change.[37]

> *Protects the health and wellbeing of citizens and protects the local economy.*

Build hard infrastructure to protect people and property from changing climate such as increased storms, floods, droughts and heatwaves.[38]

> *Protects people and property.*

Build low-carbon passively cooled buildings and retrofit air and ground heat exchange air conditioning to existing buildings.[39]

Protects people from the health effects of heatwaves.

10. UNIVERSAL BASIC INCOME

Guarantee financial payment to every citizen, unconditionally, without any obligation to work, at a level above their subsistence needs.[40, 41]

Eliminates extreme poverty.

Breaks the link between work and unsustainable consumption.

Allows citizens to retrain at any time in their career.

Allows people to choose sustainable and socially responsible jobs.

Allows citizens to think about the long term, well beyond the next payday.

Allows citizens to choose to look after the elderly and sick relatives.

UNIVERSAL BASIC INCOME COULD BREAK THE LINK BETWEEN WORK AND UNSUSTAINABLE CONSUMPTION

Increases creativity and engagement in local communities.

Reduces social stress.

Reduces healthcare costs.

Increases the number of entrepreneurs and their success rates.

Protects citizens and the economy from pandemics.

CHAPTER 9:
SAVING OUR PLANET AND OURSELVES

IMAGINE OUR FUTURE HISTORY.

CREATE THE PATHWAY.

MAKE IT HAPPEN.

The challenges of the 21st century are immense.

Do not underestimate these challenges.

We need to deal with:

  climate change,
  environmental degradation,
  global inequality and extreme poverty,
  global, national and individual security.

They must be dealt with together using win-win or win-win-win or win-win-win-win solutions.[1]

If we deal with them separately then individual solutions to one problem could make the others worse.

Remember:

  We have the technology.

  We have the resources.

WE HAVE THE TECHNOLOGY

WE HAVE THE
RESOURCES

We have the money.

We have the scientists, the entrepreneurs and the innovators.

We lack the politics and policies to make your vision of a better world happen.

So we need a plan to save our planet . . .

1. UPDATE OUR INTERNATIONAL INSTITUTIONS

Our international institutions are not fit for purpose.[2]

Many of them are over 70 years old.[2]

World Bank was formed in 1944.

United Nations was formed in 1945.

International Monetary Fund (IMF) was formed in 1945.

General Agreement on Tariffs and Trade (GATT) was signed in 1947.

GATT was replaced by the World Trade Organization (WTO) in 1995.

North Atlantic Treaty Organization (NATO) was formed in 1949.

Organization for Economic Co-operation and Development (OECD) was founded in 1961. It has 37 member countries and aims to stimulate economic progress and world trade.

G7 (or Group of Seven) was originally formed in 1975 with 6 countries: France, Germany, Italy, Japan, the United Kingdom and the United States. Canada was added in 1976. Russia was added in 1997 and then disinvited in 2014.

G20 (or Group of Twenty) consists of 19 countries and the European Union (EU). It was founded in 1999 with the aim of promoting international financial stability.

We need our international institutions to represent everyone in the world and ensure fair and equitable governance.[3]

We need a 21st-century set of Bretton Woods conferences to redesign our international governance institutions.[4]

We need to redesign the World Bank and International Monetary Fund (IMF) so they focus on developing the green sustainable economy and alleviating poverty.[5]

We need to rethink and replace the fossil fuel industrial infrastructure globally and stop the constant lobbying by companies and fossil-fuel-rich countries.[6]

We need to transform the World Trade Organization (WTO) because its fundamental goal is to ensure that trade flows as smoothly, predictably and freely as possible.[7]

WTO encourages trade and consumption and makes reductions in carbon emissions harder. It can prevent meaningful local, national and international environmental protections and regulations.[7]

We could transform the WTO into the World Sustainability Organization (WSO).[7]

We need to recognize that free trade is not always in the best interest of developing countries.[8]

Since 1960, the income gap between the Global North and Global South has roughly tripled in size.[9]

Since 1980, over $16.3tn has been transferred from the Global South to the Global North.[10]

Of this, over $4.2tn was in interest payments, direct cash transfers to big banks in New York and London.[10]

The first aim of the WSO should be to support and help restructure economies of countries that rely on fossil fuel exports.[7]

The United Nations' most powerful committee, the Security Council, consists of ten elected members and five permanent members, the winners of the Second World War.[11]

　China (population 1400 million),
　United States (330 million),
　Russian Federation (146 million),
　United Kingdom (68 million),
　France (65 million).

One country one vote in the United Nations means that citizens of the world are not equally represented.[12]

Out of the 193 countries in the UN, 75 are full democracies while 53 are fully authoritarian with little or no right of representation by citizens.[13]

Many people have little say within their own country let alone being represented at the international level.[13]

We need to rethink the United Nations so it is more representative of the world population.[12]

UN 2.0 could be more democratic, with an elected lower house.[12]

The UN Security Council could have permanent seats for the ten largest economies and rotating seats for each major region of the world.[14]

The UN Environment Agency has secondary status within the UN system. It does not have the same status as trade, health, labour or even maritime affairs, intellectual property or tourism.[15]

The UN Environment Agency's budget is small, less than a quarter of the UN World Health Organization's budget and a tenth of the UN World Food Programme's, despite being central to both these issues.[16]

We need to upgrade the UN Environment Agency to a full UN World Environment Organization (WEO).[15]

The WEO would require a budget that is at least the size of the World Health Organization's.[15]

The WEO would oversee the Sustainability Development Goals, the Convention on Biological Diversity and the Convention on Climate Change to ensure they are mutually reinforcing and not in opposition.[15]

ECONOMICS MUST FOCUS ON HUMAN WELLBEING AS THE PRIMARY MEASURE OF SUCCESS, NOT DOLLARS CREATED

2. REDESIGN ECONOMICS FOR REAL PEOPLE

After nearly forty years of neoliberalism, the IMF has declared that this approach is jeopardizing the future of the world economy.[17]

Classical economics does not work and is not suitable for the complexity and demands of our global society.[17]

Gross Domestic Product (GDP) does not measure the human condition.[18]

Economics must focus on human wellbeing as the primary measure of success, not dollars created.[19]

Wealth redistribution within and between countries needs to be addressed.[20]

Universities need to step up and become generators of wisdom and new ideas – not depositories of knowledge that simply regurgitate old economic theories.[21]

The 17 Sustainability Development Goals (SDG)[22] need to be supported:

1. No Poverty
2. Zero Hunger
3. Good Health and Well-Being
4. Quality Education
5. Gender Equality
6. Clean Water and Sanitation
7. Affordable and Clean Energy
8. Decent Work and Economic Growth
9. Industry, Innovation, and Infrastructure
10. Reduced Inequalities
11. Sustainable Cities and Communities
12. Responsible Consumption and Production
13. Climate Action
14. Life Below Water
15. Life on Land
16. Peace, Justice, and Strong Institutions
17. Partnerships for the Goals

But the SDG emphasis on economic growth needs to be revised.[23]

Post 2030 the next set of SDGs need to focus on decarbonization, energy, food and water security, removing extreme poverty and protecting the biosphere and biodiversity.[23]

3. THE NEW GEOPOLITICS

We must understand and embrace the political realities of the 21st century.

The United States must engage globally and ratify current and future international agreements, which they have up to now avoided.[24]

These include key examples of both human rights and environmental protections:[24]

- 1972 – Anti-Ballistic Missile Treaty, signed but withdrew in 2002
- 1977 – Convention on Human Rights, signed but not ratified
- 1979 – Convention on the Elimination of All Forms of Discrimination Against Women, signed but not ratified
- 1989 – Convention on the Rights of the Child, signed but not ratified
- 1990 – UN Convention on the Protection of the Rights of All Migrant Workers and Members of Their Families, not signed
- 1991 – UN Convention on the Law of the Sea, not signed
- 1992 – Rio Convention on Biological Diversity, signed but not ratified

1996 – Comprehensive Nuclear-Test-Ban Treaty, signed but
not ratified

1997 – Kyoto Protocol, signed with no intention to ratify

1997 – Mine Ban Treaty, unsigned

1998 – Setting up of the International Criminal Court,
unsigned

1999 – Convention on the Elimination of All Forms of
Discrimination Against Women, not signed

2002 – Convention Against Torture, not signed

2006 – Protection of All Persons from Enforced
Disappearance, not signed

2007 – Convention on the Rights of Persons with Disabilities,
signed but not ratified

2008 – Convention on Cluster Munitions, not signed

2011 – Anti-Counterfeiting Trade Agreement, signed but not
ratified

2013 – Arms Trade Treaty, signed but not ratified

2016 – Trans-Pacific Partnership, signed but not ratified

2017 – Paris Climate Change Agreement, signed but not
ratified and then withdrew 2020

We must acknowledge the rise of the East.[25]

The size of China's economy is rapidly catching up with the
USA's and in terms of its buying power it has already overtaken
the USA.[26]

THERE WILL BE 10 BILLION PEOPLE ON THE PLANET BY 2050

By 2050 over half the world's economy will be generated by Asia.[25]

There will be 10 billion people on the planet by 2050.[27]

5.2 billion will live in Asia.[27]

If those 5 billion consumed at the same level as today's richest billion people their consumption would be 5 to 10 times higher.[28, 29]

It would be like living on a planet of at least 46 billion people.[28, 29]

We have seen in this book the damage that our current consumption has caused – imagine the damage that could be done with 10 times the consumption.[30]

Hence any solutions to save the planet must be led or co-led by Asia.[31]

4. SET MEANINGFUL GLOBAL TARGETS

Zero global carbon emissions by 2050.[32]

Negative global carbon emissions between 2050 and 2100.[32]

ZERO GLOBAL CARBON EMISSIONS BY 2050

ZERO DEFORESTATION
BY 2030

Zero deforestation by 2030.[33]

1 trillion trees planted by 2050.[32]

Developed countries' consumption must be halved by 2050, quartered by 2100.[34]

Halve the number of people living below $5 per day by 2030.[35]

Make country-bespoke Universal Basic Income mandatory for everyone.[36]

Eliminate extreme poverty by 2050.[37]

We already have enough money in the world to do this now.[38]

We just need enlightened policies.

5. DEVELOP PATHWAYS TO OUR SHARED FUTURE

We must create positive feedbacks to change the world.

International climate change, biodiversity and environmental agreements need to be agreed as soon as possible – they legitimize action at all levels.[39]

Sustainable development must be the new core of global economics.[40]

The green economy can create jobs, livelihoods and a safer healthier environment.[40]

Country-to-country economic and political enforcement of agreements is essential to ensure everyone is committed to a better, safer future for all.[40]

International agreements and individuals' protests and behaviour change legitimize government actions on climate change and environmental protection.[40-42]

Individual consumer choice and government regulation drive companies to change their behaviour.[43, 44]

Companies taking pre-emptive moral action drive individual change and legitimize progressive government policies.[45, 46]

Governments need to build reward systems for individuals, organizations and companies that take positive action.[43]

Each country must produce its own solution that is culturally and politically appropriate.[43]

But they must address the primary global goals of reducing climate change and environmental degradation while increasing human wellbeing and security.

POSITIVES OF THE HUMAN CONDITION

We are the first species on Earth to control our own population, and the key has been women's **education** and the alleviation of poverty.[47]

In every country in the world, when women are educated to at least secondary/high school level they take control of their own fertility and the number of children they have drops.[47]

We are the first species that through our **science** can understand the consequences of our actions at local, national, regional and global scales.

We are the first species that through our **technology** can solve almost any problem we face.

THE CHALLENGE OF THE 21ST CENTURY IS THAT WE MUST LEARN TO THINK AND ACT AS A GLOBAL SPECIES

The challenge of the 21st century is that we must learn to think and act as a global species.

The next generation of humans are so interconnected they already think globally and act locally.

We must build new **political** and economic systems so that we can look after:

  all individuals;
  our global species;
  our planet –
  that we rely on for everything.

AFTERWORD

In 2020 almost everyone's lifestyle was radically altered by the Covid-19 pandemic. It also dropped our global carbon emissions through reduced transportation, reduced industrial output and reduced non-essential consumption, but not as much as you would have expected. This cessation of activity has allowed nature more room to flourish and everywhere the air was clearer and less polluted. Around the world more and more people are calling for a better, healthier and safer world post the pandemic. This is despite the call to get workers back to work, whatever the risk to their health and lives. But the genie is out of the bottle and citizens all around the world have been shown that there can be a different relationship between government, industry, civil society and our planet. A relationship where health and wellbeing are put before economic gains for a country or a small minority of individuals.

The pandemic required fast and drastic action, which had huge short-term implications for our local and global economy. It is not a model of how to deal with the long-term systemic issues of climate change and environmental degradation, but it does offer major insights into how we can collectively deal with climate change. Since the 1980s there has been a move away from government regulation and general support for society. This shift has varied between countries but can be seen everywhere. This has led to whole generations accepting the idea that markets and business know best because they are supposedly more efficient than anything organized by governments. What the Covid-19 pandemic has dramatically illustrated is that markets, business and industry – though essential for our modern life – structurally do not act in everyone's best interests. This is because companies are by their very nature focused on making profits, expanding rapidly and generating worth in terms of their share price. This focus makes companies incredibly dynamic and the private sector has played an important role in the pandemic, ensuring food supplies in the face of panic buying, retooling to produce essential medical supplies and helping to develop vaccines. But many companies have simply looked to the state for loans and bailouts. We have seen that only governments have that critical central role in maintaining our health and safety, especially in times of crisis.

While the shutdown of many sectors of the economy had environmental benefits in terms of carbon emissions and biodiversity loss, few would argue that the socio-economic costs can be justified. Moreover, it seems that the global lockdown had only a minimal effect on our carbon emissions. A recent study suggests a drop of just 7% in carbon emissions for 2020.[1] So ceasing almost all flying (approximately 100,000 flights a day) and car journeys has only a small impact on our total greenhouse gas pollution. In fact, even the greatest reduction would make 2020 global carbon emissions only as low as 2006 emissions and by 2021 emissions had jumped back to their previous levels. This is because there has been very little change in energy production during the pandemic. It is clear that green solutions are needed at a fundamental level in our economic recovery if we are to avoid the greater threat of climate change.

It is becoming clear that one of the reasons that Covid-19 is such a severe and even fatal respiratory and blood disease is that it is a zoonotic virus, a virus that has mutated, allowing it to jump from another animal into humans. This means it has a genetic signature unknown to our immune systems, delaying our ability to develop antibodies that can fight the infection. It is likely the illegal trade in endangered animals such as bats and pangolins through inhumane 'wet markets' in China and South East Asia enabled the virus to jump into humans. The extremely high risks of such zoonotic virus outbreaks were indicated by previous outbreaks, such as the HN51 virus in 1996 and the SARS outbreak in 2002–3. After each of these the Chinese government banned all wet markets and related wildlife trade but then for cultural reasons relaxed all the restrictions.

A better governance approach would be to heed these zoonotic virus warnings from experts and take a strategic approach to reducing this trade through proactive measures by all governments, such as promoting cultural change, along with gradual regulatory restrictions. At this stage, the emphasis should be on ensuring a commitment to a long-term ban on the wildlife trade through a coordinated and strategic approach, especially given the global impacts on many rare species of growing demand to feed the wildlife trade. The very health and safety of many people will rely on us changing our relationship with nature and ensuring protection to biodiversity and unique ecosystems around the world.

Throughout this book I hope I have shown that government incentives, policies, nudges, taxation, regulation and enforcement can shape our society to ensure the best outcomes. So post Covid-19 we need to harness this new dynamic, caring role of governments and use this to shift national and global economies to a more sustainable footprint. If done properly, it will also help us deal with the next pandemic and it may even help prevent another pandemic happening, if we change the governance of nature and wildlife. For example, shifting to localized renewable energy increases energy and job security, essential when dealing with a global pandemic lockdown. Removing fossil fuel subsidies provides trillions of extra dollars for our healthcare systems. Dramatically decreasing meat consumption, increasing animal welfare and fully protecting wildlife and biodiversity increases human health and reduces the chance of zoonosis and thus new pandemics developing.

Instigating Universal Basic Income has been shown to reduce non-essential consumption, and lowers people's carbon footprint and protects the economy from the next pandemic as everyone will have enough money to live on however long social distancing must be maintained.

Covid-19 has changed our view of governments and their role in society. By embracing this, we can now ensure that win–win solutions are adopted to deal with the climate change and environmental emergency. Our role as concerned citizens is to make sure governments do act in the interest of everyone – that they do protect our planet's biodiversity, precious resources and help stabilize our climate. We the people are the nation and our governments must govern for us.

For my last birthday, a friend sent me a card with this written on the front:

'The future awaits those with the courage to create it.'

I hope this book shows that all of us, if we have the courage, can create a better, fairer, safer future for all.

REFERENCES

CHAPTER 1: HISTORY OF OUR PLANET

1. Planck Collaboration, 'Planck 2018 results, VI. Cosmological parameters', *Astronomy & Astrophysics* (2018), 1–72.
2. G. Lemaître, 'The beginning of the world from the point of view of quantum theory', *Nature* 127 (1931), 706.
3. E. Hubble, 'A relation between distance and radial velocity among extra-galactic nebulae', *Proceedings of the National Academy of Sciences of the USA* 15 (1929), 168–73.
4. A. A. Penzias & R. W. Wilson, 'A measurement of excess antenna temperature at 4080 Mc/s', *The Astrophysical Journal* 142 (1965), 419–21.
5. I. Ferreras, *Fundamentals of Galaxy Dynamics, Formation and Evolution* (UCL Press, 2019).
6. S. W. Stahler & F. Palla, *The Formation of Stars* (Wiley, 2004).
7. A. S. Eddington, 'The internal constitution of the stars', *Science* 52 (1920), 233–40.
8. F. Hoyle, 'On nuclear reactions occurring in very hot stars. I. The synthesis of elements from carbon to nickel', *The Astrophysical Journal Supplement Series* 1 (1954), 121–46.
9. D. A. Fischer & J. Valenti, 'The planet-metallicity correlation', *The Astrophysical Journal* 622 (2005), 1102–17.
10. E. M. Burbidge, G. R. Burbidge, W. A. Fowler & F. Hoyle, 'Synthesis of the elements in stars', *Reviews of Modern Physics* 29 (1957), 547–650.

REFERENCES

11. A. Bouvier & M. Wadhwa, 'The age of the Solar System redefined by the oldest Pb-Pb age of a meteoritic inclusion', *Nature Geoscience* 3 (2010), 637–41.

12. S. Jain, 'The Cosmic Bodies' in *Fundamentals of Physical Geology*, ed. S. Jain (Springer India, 2014), pp. 37–53.

13. G. Faure & T. M. Mensing, 'From Speculation to Understanding' in *Introduction to Planetary Science: The Geological Perspective*, ed. G. Faure & T. M. Mensing (Springer Netherlands, 2007), pp. 13–21.

14. C. H. Langmuir & W. Broecker, *How to Build a Habitable Planet: The Story of Earth from the Big Bang to Humankind* (Princeton University Press, 2012).

15. J. Laskar, F. Joutel & P. Robutel, 'Stabilization of the Earth's obliquity by the Moon', *Nature* 361 (1993), 615–17.

16. M. S. Dodd *et al.*, 'Evidence for early life in Earth's oldest hydrothermal vent precipitates', *Nature* 543 (2017), 60–64.

17. M. McFall-Ngai *et al.*, 'Animals in a bacterial world, a new imperative for the life sciences', *Proceedings of the National Academy of Sciences of the USA* 110 (2013), 3229–36.

18. A. H. Knoll, E. J. Javaux, D. Hewitt & P. Cohen, 'Eukaryotic organisms in Proterozoic oceans', *Philosophical Transactions of the Royal Society B: Biological Sciences* 361 (2006), 1023–38.

19. L. Chen, S. Xiao, K. Pang, C. Zhou & X. Yuan, 'Cell differentiation and germ-soma separation in Ediacaran animal embryo-like fossils', *Nature* 516 (2014), 238–41.

20. A. C. Maloof *et al.*, 'The earliest Cambrian record of animals and ocean geochemical change', *GSA Bulletin* 122 (2010), 1731–74.

21. D. G. Shu *et al.*, 'Lower Cambrian vertebrates from south China', *Nature* 402 (1999), 42–6.

22. S. G. Lucas & Z. Luo, '*Adelobasileus* from the Upper Triassic of West Texas: The oldest mammal', *Journal of Vertebrate Paleontology* 13 (1993), 309–34.

23. M. Maslin, *The Cradle of Humanity: How the Changing Landscape of Africa Made Us So Smart* (Oxford University Press, 2017).

24. J. H. Kaas, 'The origin and evolution of neocortex: From early mammals to modern humans', *Progress in Brain Research* 250 (2019), 61–81.

REFERENCES

25. T. J. D. Halliday, P. Upchurch & A. Goswami, 'Eutherians experienced elevated evolutionary rates in the immediate aftermath of the Cretaceous–Palaeogene mass extinction', *Proceedings of the Royal Society B: Biological Sciences* 283 (2016), 1–8.

26. P. R. Renne *et al.*, 'Time scales of critical events around the Cretaceous–Paleogene boundary', *Science* 339 (2013), 684–7.

27. C. A. Emerling, F. Delsuc & M. W. Nachman, 'Chitinase genes (CHIAs) provide genomic footprints of a post-Cretaceous dietary radiation in placental mammals', *Science Advances* 4 (2018), 1–10.

28. J. D. Archibald, *Extinction and Radiation: How the Fall of Dinosaurs Led to the Rise of Mammals* (Johns Hopkins University Press, 2011).

29. X. Ni *et al.*, 'The oldest known primate skeleton and early haplorhine evolution', *Nature* 498 (2013), 60–64.

30. S. Shultz, C. Opie & Q. D. Atkinson, 'Stepwise evolution of stable sociality in primates', *Nature* 479 (2011), 219–22.

31. C. V. Ward, A. S. Hammond, J. M. Plavcan & D. R. Begun, 'A late Miocene hominid partial pelvis from Hungary', *Journal of Human Evolution* 136 (2019), 1–25.

32. M. Brunet *et al.*, 'A new hominid from the Upper Miocene of Chad, Central Africa', *Nature* 418 (2002), 145–51.

33. B. Senut *et al.*, 'Palaeoenvironments and the origin of hominid bipedalism', *Historic Biology* 30 (2018), 284–96.

34. S. Harmand *et al.*, '3.3-million-year-old stone tools from Lomekwi 3, West Turkana, Kenya', *Nature* 521 (2015), 310–15.

35. S. Shultz & M. Maslin, 'Early human speciation, brain expansion and dispersal influenced by African climate pulses', *PLOS ONE* 8 (2013), 1–7; and M. A. Maslin *et al.*, 'A synthesis of the theories and concepts of early human evolution', *Philosophical Transactions of the Royal Society B: Biological Sciences* 370 (2015), 20140064: https://doi.org/10.1098/rstb.2014.0064.

36. F. Carotenuto *et al.*, 'Venturing out safely: The biogeography of *Homo erectus* dispersal out of Africa', *Journal of Human Evolution* 95 (2016), 1–12.

37. N. T. Roach, M. Venkadesan, M. J. Rainbow & D. E. Lieberman, 'Elastic energy storage in the shoulder and the evolution of high-speed throwing in *Homo*', *Nature* 498 (2013), 483–6.

38. L. C. Aiello & C. Key, 'Energetic consequences of being a *Homo erectus* female', *American Journal of Human Biology* 14 (2002), 551–65.

39. C. A. O'Connell & J. M. DeSilva, 'Mojokerto revisited: Evidence for an intermediate pattern of brain growth in *Homo erectus*', *Journal of Human Evolution* 65 (2013), 156–61.

40. D. M. Bramble & D. E. Lieberman, 'Endurance running and the evolution of *Homo*', *Nature* 432 (2004), 345–52.

41. S. W. Simpson *et al.*, 'A Female *Homo erectus* pelvis from Gona, Ethiopia', *Science* 322 (2008), 1089–92.

42. J. C. A. Joordens *et al.*, '*Homo erectus* at Trinil on Java used shells for tool production and engraving', *Nature* 518 (2015), 228–31.

43. K. G. Hatala *et al.*, 'Footprints reveal direct evidence of group behavior and locomotion in *Homo erectus*', *Scientific Reports* 6 (2016), 1–9.

44. J. A. J. Gowlett & R. W. Wrangham, 'Earliest fire in Africa: Towards the convergence of archaeological evidence and the cooking hypothesis', *Azania: Archaeological Research in Africa* 48 (2013), 5–30.

45. D. Richter *et al.*, 'The age of the hominin fossils from Jebel Irhoud, Morocco, and the origins of the Middle Stone Age', *Nature* 546 (2017), 293–6.

46. A. Bouzouggar *et al.*, '82, 000-year-old shell beads from North Africa and implications for the origins of modern human behavior', *Proceedings of the National Academy of Sciences of the USA* 104 (2007), 9964–9.

47. R. Nielsen *et al.*, 'Tracing the peopling of the world through genomics', *Nature* 541 (2017), 302–10.

48. B. Hood, *The Domesticated Brain: A Pelican Introduction* (Penguin, 2014), p. 336.

49. J. F. Hoffecker, *Modern Humans: Their African Origin and Global Dispersal* (Columbia University Press, 2017).

50. C. Gamble, J. Gowlett & R. Dunbar, *Thinking Big: How the Evolution of Social Life Shaped the Human Mind* (Thames & Hudson, 2014), p. 224.

51.   S. L. Lewis & M. A. Maslin, *The Human Planet: How We Created the Anthropocene* (Penguin, 2018).

CHAPTER 2: HISTORY OF HUMANITY

1.   J. F. Hoffecker, *Modern Humans: Their African Origin and Global Dispersal* (Columbia University Press, 2017).

2.   M. W. Pedersen *et al.*, 'Postglacial viability and colonization in North America's ice-free corridor', *Nature* 537 (2016), 45–9.

3.   S. L. Lewis & M. A. Maslin, *The Human Planet: How We Created the Anthropocene* (Penguin, 2018).

4.   A. D. Barnosky, P. L. Koch, R. S. Feranec, S. L. Wing & A. B. Shabel, 'Assessing the causes of Late Pleistocene extinctions on the continents', *Science* 306 (2004), 70–75.

5.   A. D. Barnosky, 'Megafauna biomass tradeoff as a driver of Quaternary and future extinctions', *Proceedings of the National Academy of Sciences of the USA* 105 (2008), 11543–8.

6.   M. Maslin, *The Cradle of Humanity: How the Changing Landscape of Africa Made Us So Smart* (Oxford University Press, 2017).

7.   Y. Malhi, 'The Metabolism of a Human-Dominated Planet' in *Is the Planet Full?*, ed. I. Goldin (Oxford University Press, 2014), pp. 142–63.

8.   J. M. Diamond, *Guns, Germs, and Steel: The Fates of Human Societies* (W. W. Norton & Co., 1999).

9.   G. Larson *et al.*, 'Current perspectives and the future of domestication studies', *Proceedings of the National Academy of Sciences of the USA* 111 (2014), 6139–46.

10.   N. D. Wolfe, C. P. Dunavan & J. Diamond, 'Origins of major human infectious diseases', *Nature* 447 (2007), 279–83.

11.   W. F. Ruddiman, 'The anthropogenic greenhouse era began thousands of years ago', *Climatic Change* 61 (2003), 261–93.

12.   W. F. Ruddiman, Z. Guo, X. Zhou, H. Wu & Y. Yu, 'Early rice farming and anomalous methane trends', *Quaternary Science Reviews* 27 (2008), 1291–5.

13. W. F. Ruddiman *et al.*, 'Late Holocene climate: Natural or anthropogenic?', *Reviews of Geophysics* 54 (2016), 93–118.

14. US Census Bureau, 'Historical estimates of world population' (2018): https://www.census.gov/data/tables/time-series/demo/international-programs/historical-est-worldpop.html.

15. R. H. Fuson, *The Log of Christopher Columbus* (International Marine Publishing Company, 1987).

16. A. Koch, C. Brierley, M. A. Maslin & S. L. Lewis, 'Earth system impacts of the European arrival and Great Dying in the Americas after 1492', *Quaternary Science Reviews* 207 (2019), 13–36.

17. A. W. Crosby, 'Virgin soil epidemics as a factor in the aboriginal depopulation in America', *The William and Mary Quarterly* 33 (1976), 289–99.

18. R. S. Walker, L. Sattenspiel & K. R. Hill, 'Mortality from contact-related epidemics among indigenous populations in Greater Amazonia', *Scientific Reports* 5 (2015), 1–9.

19. H. F. Dobyns, 'Disease transfer at contact', *Annual Review of Anthropology* 22 (1993), 273–91.

20. N. D. Cook, *Born to Die: Disease and New World Conquest, 1492–1650* (Cambridge University Press, 1998).

21. C. C. Mann, *1491: New Revelations of the Americas before Columbus* (Knopf, 2005).

22. A. W. Crosby, *The Columbian Exchange: Biological and Cultural Consequences of 1492* (Praeger, 2003).

23. D. Wootton, *The Invention of Science: A New History of the Scientific Revolution* (Penguin, 2015).

24. I. Wallerstein, 'The rise and future demise of the world capitalist system: Concepts for comparative analysis', *Comparative Studies in Society and History* 16 (1974), 387–415.

25. H. S. Klein, *The Atlantic Slave Trade* (Cambridge University Press, 1999).

26. J. Micklethwait & A. Wooldridge, *The Company: A Short History of a Revolutionary Idea* (Phoenix, 2005).

27. G. J. Ames, *The Globe Encompassed: The Age of European Discovery, 1500–1700* (Pearson Prentice Hall, 2008).

28. W. Dalrymple, *The Anarchy: The Relentless Rise of the East India Company* (Bloomsbury, 2019).

29. J. Horn, L. N. Rosenband & M. R. Smith, *Reconceptualizing the Industrial Revolution* (MIT Press, 2010).

30. R. C. Allen, *The British Industrial Revolution in Global Perspective* (Cambridge University Press, 2009).

31. UN DESA, Population Division, 'The 1998 revision of the United Nations population projections', *Population and Development Review* 24 (1998), 891–5.

32. B. W. Clapp, *An Environmental History of Britain since the Industrial Revolution* (Taylor & Francis, 1994).

33. IPCC, *Climate Change 2013: The Physical Science Basis. Contribution of Working Group I to the Fifth Assessment Report of the Intergovernmental Panel on Climate Change* (2013): https://www.ipcc.ch/site/assets/uploads/2018/02/WG1AR5_all_final.pdf.

34. C. I. Archer, J. R. Ferris, H. H. Herwig & T. H. E. Travers, *World History of Warfare* (University of Nebraska Press, 2002).

35. G. O'Reilly, *Aligning Geopolitics, Humanitarian Action and Geography in Times of Conflict* (Springer Nature, 2019).

36. H. G. Wells, *The War That Will End War* (Duffield & Co., 1914).

37. P. N. Stearns, *The Industrial Revolution in World History* (Taylor & Francis, 2013).

38. A. Klasen (ed.), *The Handbook of Global Trade Policy* (Wiley-Blackwell, 2020).

39. W. Steffen, W. Broadgate, L. Deutsch, O. Gaffney & C. Ludwig, 'The trajectory of the Anthropocene: The Great Acceleration', *The Anthropocene Review* 2 (2015), 81–98.

40. N. Lewkowicz, *The United States, the Soviet Union and the Geopolitical Implications of the Origins of the Cold War* (Anthem Press, 2018).

41. R. S. Norris & H. M. Kristensen, 'Global nuclear weapons inventories, 1945–2010', *Bulletin of the Atomic Scientists* 66 (2010), 77–83.

42. J. R. McNeill & P. Engelke, *The Great Acceleration: An Environmental History of the Anthropocene since 1945* (Harvard University Press, 2014).

43. United Nations, *World Population Prospects 2019: Highlights* (2019): https://population.un.org/wpp/Publications/Files/WPP2019_Highlights.pdf.

44. J. H. Brown *et al.*, 'Energetic limits to economic growth', *BioScience* 61 (2011), 19–26.

45. D. McNally, *Global Slump: The Economics and Politics of Crisis and Resistance* (PM Press, 2011).

46. R. Shretta, 'The economic impact of COVID-19' (University of Oxford, 7 April 2020): https://www.research.ox.ac.uk/Article/2020-04-07-the-economic-impact-of-covid-19.

CHAPTER 3: STATE OF OUR WORLD

1. S. L. Lewis & M. A. Maslin, *The Human Planet: How We Created the Anthropocene* (Penguin, 2018).

2. C. N. Waters *et al.*, 'The Anthropocene is functionally and stratigraphically distinct from the Holocene', *Science* 351 (2016), 137–47.

3. B. H. Wilkinson, 'Humans as geologic agents: A deep-time perspective', *Geology* 33 (2005), 161–4.

4. R. M. Hazen, E. S. Grew, M. J. Origlieri & R. T. Downs, 'On the mineralogy of the "Anthropocene Epoch"', *American Mineralogist* 102 (2017), 595–611.

5. W. Steffen, W. Broadgate, L. Deutsch, O. Gaffney & C. Ludwig, 'The trajectory of the Anthropocene: The Great Acceleration', *The Anthropocene Review* 2 (2015), 81–98.

6. J. Boucher & D. Friot, *Primary Microplastics in the Oceans: A Global Evaluation of Sources* (IUCN, 2017): https://www.iucn.org/content/primary-microplastics-oceans.

7. BBC News, 'Mariana Trench: Deepest-ever sub dive finds plastic bag' (13 May 2019): https://www.bbc.co.uk/news/science-environment-48230157.

REFERENCES

8. J. Chapman, 'Introduction' in *Routledge Handbook of Sustainable Product Design*, ed. J. Chapman (Taylor & Francis, 2017).

9. J. N. Galloway *et al.*, 'Transformation of the nitrogen cycle: Recent trends, questions, and potential solutions', *Science* 320 (2008), 889–92.

10. United Nations, *World Population Prospects 2019: Highlights* (2019): https://population.un.org/wpp/Publications/Files/WPP2019_Highlights.pdf.

11. T. W. Crowther *et al.*, 'Mapping tree density at a global scale', *Nature* 525 (2015), 201–5.

12. V. Smil, *Harvesting the Biosphere: What We Have Taken from Nature* (MIT Press, 2013).

13. FAO, *The State of World Fisheries and Aquaculture 2018: Meeting the Sustainable Development Goals* (2018): http://www.fao.org/3/I9540EN/i9540en.pdf.

14. FAO, *World Food and Agriculture: Statistical Pocketbook 2019* (2019): http://www.fao.org/3/ca6463en/CA6463EN.pdf.

15. J. Baillie *et al.*, *2004 IUCN Red List of Threatened Species: A Global Species Assessment* (IUCN, 2004).

16. L. Mitchell, E. Brook, J. E. Lee, C. Buizert & T. Sowers, 'Constraints on the late Holocene anthropogenic contribution to the atmospheric methane budget', *Science* 342 (2013), 964–6.

17. WMO, *WMO Greenhouse Gas Bulletin No. 15: The State of Greenhouse Gases in the Atmosphere Based on Global Observations through 2018* (2018): https://library.wmo.int/doc_num.php?explnum_id=10100.

18. IPCC, *Climate Change 2013: The Physical Science Basis. Contribution of Working Group I to the Fifth Assessment Report of the Intergovernmental Panel on Climate Change* (2013): https://www.ipcc.ch/site/assets/uploads/2018/02/WG1AR5_all_final.pdf.

19. H. Ritchie & M. Roser, 'CO_2 and greenhouse gas emissions' (Our World in Data, 2017): https://ourworldindata.org/co2-and-other-greenhouse-gas-emissions.

20. M. Willeit, A. Ganopolski, R. Calov & V. Brovkin, 'Mid-Pleistocene transition in glacial cycles explained by declining CO_2 and regolith removal', *Science Advances* 5 (2019), 1–8.

REFERENCES

21. M. Maslin, *Climate Change: A Very Short Introduction* (Oxford University Press, 2014).

22. K. Caldeira & M. E. Wickett, 'Anthropogenic carbon and ocean pH', *Nature* 425 (2003), 365.

23. IUCN, *Ocean Deoxygenation: Everyone's Problem – Causes, Impacts, Consequences and Solutions* (2019): https://portals.iucn.org/library/sites/library/files/documents/2019-048-En.pdf.

24. J. Fourier, 'Remarques générales sur les températures du globe terrestre et des espaces planétaires', *Annales de Chimie et de Physique* 27 (1824), 136–67.

25. IPCC, *Climate Change 2014: Synthesis Report. Contribution of Working Groups I, II and III to the Fifth Assessment Report of the Intergovernmental Panel on Climate Change* (2014): https://www.ipcc.ch/site/assets/uploads/2018/02/SYR_AR5_FINAL_full.pdf.

26. J. Wang *et al.*, 'Global land surface air temperature dynamics since 1880', *International Journal of Climatology* 38 (2018), 466–74.

27. L. Cheng *et al.*, 'Record-setting ocean warmth continued in 2019', *Advances in Atmospheric Sciences* 37 (2020), 137–42.

28. R. S. Nerem *et al.*, 'Climate-change-driven accelerated sea-level rise detected in the altimeter era', *Proceedings of the National Academy of Sciences of the USA* 115 (2018), 2022–5.

29. K. E. Kunkel *et al.*, 'Trends and extremes in Northern Hemisphere snow characteristics', *Current Climate Change Reports* 2 (2016), 65–73.

30. J. Stroeve & D. Notz, 'Changing state of Arctic sea ice across all seasons', *Environmental Research Letters* 13 (2018), 1–23.

31. World Glacier Monitoring Service, *Global Glacier Change Bulletin No. 2 (2014–2015)* (2017): https://wgms.ch/downloads/WGMS_GGCB_02.pdf.

32. A. Shepherd *et al.*, 'Mass balance of the Greenland Ice Sheet from 1992 to 2018', *Nature* 579 (2019): doi:10.1038/s41586-019-1855-2.

33. E. Rignot *et al.*, 'Four decades of Antarctic Ice Sheet mass balance from 1979–2017', *Proceedings of the National Academy of Sciences of the USA* 116 (2019), 1095–103.

34. B. K. Biskaborn *et al.*, 'Permafrost is warming at a global scale', *Nature Communications* 10 (2019), 1–11.

35. S. Piao *et al.*, 'Plant phenology and global climate change: Current progresses and challenges', *Global Change Biology* 25 (2019), 1922–40.

36. C. Howard *et al.*, 'Flight range, fuel load and the impact of climate change on the journeys of migrant birds', *Proceedings of the Royal Society B: Biological Sciences* 285 (2018), 1–9.

37. C. Parmesan, 'Ecological and evolutionary responses to recent climate change', *Annual Review of Ecology, Evolution, and Systematics* 37 (2006), 637–69.

38. K. T. Bhatia *et al.*, 'Recent increases in tropical cyclone intensification rates', *Nature Communications* 10 (2019), 1–9.

39. S. A. Kulp & B. H. Strauss, 'New elevation data triple estimates of global vulnerability to sea-level rise and coastal flooding', *Nature Communications* 10 (2019), 1–12.

40. E. Bevacqua *et al.*, 'Higher probability of compound flooding from precipitation and storm surge in Europe under anthropogenic climate change', *Science Advances* 5 (2019), 1–7.

41. B. I. Cook, J. S. Mankin & K. J. Anchukaitis, 'Climate change and drought: From past to future', *Current Climate Change Reports* 4 (2018), 164–79.

42. P. Pfleiderer, C.-F. Schleussner, K. Kornhuber & D. Coumou, 'Summer weather becomes more persistent in a 2°C world', *Nature Climate Change* 9 (2019), 666–71.

43. M. W. Jones *et al.*, 'Climate change increases the risk of wildfires' (Science Brief, 2020): https://sciencebrief.org/briefs/wildfires.

44. NOAA, 'State of the Climate: Global Climate Report for Annual 2019' (2020): https://www.ncdc.noaa.gov/sotc/global/201913.

45. Oxfam, *Extreme Carbon Inequality* (2015): https://oi-files-d8-prod.s3.eu-west-2.amazonaws.com/s3fs-public/file_attachments/mb-extreme-carbon-inequality-021215-en.pdf.

46. World Bank, *Atlas of Sustainable Development Goals 2018: From World Development Indicators* (2018): http://documents.worldbank.org/curated/en/590681527864542864/pdf/126797-PUB-PUBLIC.pdf.

47. FAO, *The State of Food Security and Nutrition in the World 2019: Safeguarding Against Economic Slowdowns and Downturns* (2019): http://www.fao.org/3/ca5162en/ca5162en.pdf.

48. US Department of Agriculture, 'Food prices and spending' (2019): https://www.ers.usda.gov/data-products/ag-and-food-statistics-charting-the-essentials/food-prices-and-spending/.

49. FAO of the United Nations, *Building on Gender, Agrobiodiversity and Local Knowledge* (2005): http://www.fao.org/3/a-y5956e.pdf.

50. E. Holt-Giménez, A. Shattuck, M. Altieri, H. Herren & S. Gliessman, 'We already grow enough food for 10 billion people . . . and still can't end hunger', *Journal of Sustainable Agriculture* 36 (2012), 595–8.

51. UNICEF, *Levels and Trends in Child Mortality: Report 2019* (2019): https://www.unicef.org/sites/default/files/2019-10/UN-IGME-child-mortality-report-2019.pdf.

52. J. Buzby, H. Wells & J. Hyman, *The Estimated Amount, Value, and Calories of Postharvest Food Losses at the Retail and Consumer Levels in the United States* (US Department of Agriculture, 2014): https://www.ers.usda.gov/webdocs/publications/43833/43680_eib121.pdf.

53. FAO, *The State of Food and Agriculture: Innovation in Family Farming* (2014): http://www.fao.org/3/a-i4040e.pdf; doi:10.13140/2.1.3919.7765.

54. World Bank, 'Employment in agriculture (% of total employment)' (2019): https://data.worldbank.org/indicator/SL.AGR.EMPL.ZS.

55. IEA, IRENA, UNSD, WB, WHO, *Tracking SDG 7: The Energy Progress Report 2019* (2019): https://trackingsdg7.esmap.org/data/files/download-documents/2019-Tracking%20SDG7-Full%20Report.pdf.

CHAPTER 4: TAXONOMY OF DENIAL

1. N. McCarthy, 'Oil and gas giants spend millions lobbying to block climate change policies', *Forbes* (25 March 2019): https://www.forbes.com/sites/niallmccarthy/2019/03/25/oil-and-gas-giants-spend-millions-lobbying-to-block-climate-change-policies/#1bddde527c4f.

2. N. Stern *et al.*, *Stern Review: The Economics of Climate Change* (UK Government, 2006).

3. IPCC, *Climate Change 2021: The Physical Science Basis. Contribution of Working Group I to the Sixth Assessment Report of the Intergovernmental Panel on Climate Change* (2021).

4. R. Neukom, N. Steiger, J. J. Gómez-Navarro *et al.*, 'No evidence for globally coherent warm and cold periods over the preindustrial Common Era', *Nature* 571 (2019), 550–54: https://doi.org/10.1038/s41586-019-1401-2.

5. N. Watts *et al.*, 'The 2020 Report of The *Lancet* Countdown on Health and Climate Change', *Lancet* 396 (2020): 1129–306: https://www.thelancet.com/countdown-health-climate.

6. NASA-JPL, 'Graphic: Temperature vs solar activity' (2020): https://climate.nasa.gov/climate_resources/189/graphic-temperature-vs-solar-activity/.

7. T. Sloan & A. W. Wolfendale, 'Cosmic rays, solar activity and the climate', *Environmental Research Letters* 8 (2013), 1–7.

8. E. Foote, 'Circumstances affecting the heat of sun's rays', *American Journal of Art and Science* XXII (1856), 382–3.

9. WHO, *Concise International Chemical Assessment Document 61, Hydrogen Cyanide and Cyanides: Human Health Aspects* (2004), p. 73: https://www.who.int/ipcs/publications/cicad/en/cicad61.pdf.

10. IPCC, *Climate Change 2013: The Physical Science Basis. Contribution of Working Group I to the Fifth Assessment Report of the Intergovernmental Panel on Climate Change* (2013): https://www.ipcc.ch/site/assets/uploads/2018/02/WG1AR5_all_final.pdf.

11. NASA-JPL, 'Scientific consensus: Earth's climate is warming' (2020): https://climate.nasa.gov/scientific-consensus/.

12. M. Maslin, 'Cascading uncertainty in climate change models and its implications for policy', *Geographical Journal* 179 (2013), 264–71.

13. I. Ghosh, 'The $88 trillion world economy in one chart', *Visual Capitalist* (14 September 2020): https://www.visualcapitalist.com/the-88-trillion-world-economy-in-one-chart/.

14. New Climate Economy, *Unlocking the Inclusive Growth Story of the 21st Century: Accelerating Climate Action in Urgent Times* (2018): https://newclimateeconomy.report/2018/wp-content/uploads/sites/6/2019/04/NCE_2018Report_Full_FINAL.pdf.

15. D. Coady, I. Parry, N. Le & B. Shang, *Global Fossil Fuel Subsidies Remain Large: An Update Based on Country-Level Estimates* (IMF working paper, 2019): https://www.imf.org/en/Publications/WP/Issues/2019/05/02/Global-Fossil-Fuel-Subsidies-Remain-Large-An-Update-Based-on-Country-Level-Estimates-46509.

16. S. A. Kulp & B. H. Strauss, 'New elevation data triple estimates of global vulnerability to sea-level rise and coastal flooding', *Nature Communications* 10 (2019), 1–12.

17. E. Bevacqua *et al.*, 'Higher probability of compound flooding from precipitation and storm surge in Europe under anthropogenic climate change', *Science Advances* 5 (2019), 1–7.

18. B. I. Cook, J. S. Mankin & K. J. Anchukaitis, 'Climate change and drought: From past to future', *Current Climate Change Reports* 4 (2018), 164–79.

19. P. Pfleiderer, C.-F. Schleussner, K. Kornhuber & D. Coumou, 'Summer weather becomes more persistent in a 2°C world', *Nature Climate Change* 9 (2019), 666–71.

20. D. Shaposhnikov *et al.*, 'Mortality related to air pollution with the Moscow heat wave and wildfire of 2010', *Epidemiology* 25 (2014), 359–64.

21. D. Barriopedro, E. M. Fischer, J. Luterbacher, R. M. Trigo & R. García-Herrera, 'The hot summer of 2010: Redrawing the temperature record map of Europe', *Science* 332 (2011), 220–24.

22. J. M. McGrath & D. B. Lobell, 'Regional disparities in the CO_2 fertilization effect and implications for crop yields', *Environmental Research Letters* 8 (2013): https://iopscience.iop.org/article/10.1088/1748-9326/8/1/014054.

23. M. Maslin, *Climate Change: A Very Short Introduction* (Oxford University Press, 2021).

24. NOAA, Science on a Sphere, 'Ocean–atmosphere CO_2 exchange' (2020): https://sos.noaa.gov/datasets/ocean-atmosphere-co2-exchange/.

25. NYDF Assessment Partners, *Protecting and Restoring Forests: A Story of Large Commitments yet Limited Progress* (2019): https://forestdeclaration.org/images/uploads/resource/2019NYDFReport.pdf.

26. P. Guertler & P. Smith, *Cold Homes and Excess Winter Deaths: A Preventable Public Health Epidemic That Can No Longer Be Tolerated* (2018): https://www.e3g.org/publications/cold-homes-and-excess-winter-deaths-a-preventable-public-health-epidemic/.

27. NOAA, US National Weather Service, 'Weather related fatality and injury statistics: Weather fatalities 2019' (2019): https://www.weather.gov/hazstat/.

28. WHO, *Public Health Advice on Preventing Health Effects of Heat* (2011): http://www.euro.who.int/__data/assets/pdf_file/0007/147265/Heat_information_sheet.pdf?ua=1.

29. J. Tollefson, 'The hard truths of climate change – by the numbers', *Nature Briefing* 573 (2019), 325–7.

30. M. Rocha *et al.*, *Historical Responsibility for Climate Change – From Countries Emissions to Contribution to Temperature Increase* (Climate Analytics, 2015): https://www.climateanalytics.org/media/historical_responsibility_report_nov_2015.pdf.

31. Energy & Climate Intelligence Unit, *Net Zero: Why?* (2018): https://ca1-eci.edcdn.com/briefings-documents/net-zero-why-PDF-compressed.pdf?mtime=20190529123722.

32. L. E. Erickson & M. Jennings, 'Energy, transportation, air quality, climate change, health nexus: Sustainable energy is good for our health', *AIMS Public Health* 4 (2017), 47–61.

33. L. Georgeson & M. Maslin, 'Estimating the scale of the US green economy within the global context', *Palgrave Communications* 5 (2019), 121.

34. M. Maslin & S. Lewis, 'Reforesting an area the size of the US needed to help avert climate breakdown, say researchers – are they right?' *The Conversation* (4 July 2019): https://theconversation.com/reforesting-an-area-the-size-of-the-us-needed-to-help-avert-climate-breakdown-say-researchers-are-they-right-119842.

CHAPTER 5: POTENTIAL FUTURES – NIGHTMARE OR ECOTOPIA?

1. M. Collins *et al.*, 'Long-term Climate Change: Projections, Commitments and Irreversibility' in *Climate Change 2013: The Physical Science Basis. Contribution of Working Group I to the Fifth Assessment Report of the Intergovernmental Panel on Climate Change* (2013): https://www.ipcc.ch/site/assets/uploads/2018/02/WG1AR5_all_final.pdf.

2. P. Pfleiderer, C.-F. Schleussner, K. Kornhuber & D. Coumou, 'Summer weather becomes more persistent in a 2°C world', *Nature Climate Change* 9 (2019), 666–71.

3. ILO, *Working on a Warmer Planet: The Impact of Heat Stress on Labour Productivity and Decent Work* (2019): https://www.ilo.org/wcmsp5/groups/public/---dgreports/---dcomm/---publ/documents/publication/wcms_711919.pdf.

4. M. A. Krawchuk, M. A. Moritz, M.-A. Parisien, J. Van Dorn & K. Hayhoe, 'Global pyrogeography: The current and future distribution of wildfire', *PLOS ONE* 4 (2009), 1–12.

5. H. Morita & P. Kinney, 'Wildfires, Air Pollution, Climate Change and Health' in *Climate Change and Global Health*, ed. C. D. Butler (CABI, 2014), pp. 114–23.

6. D. Laffoley & J. M. Baxter, *Explaining Ocean Warming: Causes, Scale, Effects and Consequences* (IUCN, 2016): https://portals.iucn.org/library/node/46254.

REFERENCES

7. Great Barrier Reef Marine Park Authority, *Great Barrier Reef Outlook Report 2019: In Brief* (2019): http://elibrary.gbrmpa.gov.au/jspui/bitstream/11017/3478/1/Outlook-In-Brief-2019.pdf.

8. B. I. Cook, J. S. Mankin & K. J. Anchukaitis, 'Climate change and drought: From past to future', *Current Climate Change Reports* 4 (2018), 164–79.

9. A. Mirzabaev *et al.*, 'Desertification' in *Climate Change and Land: an IPCC Special Report on Climate Change, Desertification, Land Degradation, Sustainable Land Management, Food Security, and Greenhouse Gas Fluxes in Terrestrial Ecosystems* (2019): https://www.ipcc.ch/srccl/.

10. R. Engelman & P. LeRoy, *Sustaining Water: Population and the Future of Renewable Water Supplies* (Population Action International, 1993).

11. FAO, *Climate Change and Food Security: Risks and Responses* (2016): http://www.fao.org/3/a-i5188e.pdf.

12. D. Notz & J. Stroeve, 'The trajectory towards a seasonally ice-free Arctic Ocean', *Current Climate Change Reports* 4 (2018), 407–16.

13. M. Meredith *et al.*, 'Polar Regions' in *IPCC 2019: Special Report on the Ocean and Cryosphere in a Changing Climate* (2019): https://www.ipcc.ch/site/assets/uploads/sites/3/2019/11/07_SROCC_Ch03_FINAL.pdf.

14. R. Hock *et al.*, 'High Mountain Areas' in *IPCC 2019: Special Report on the Ocean and Cryosphere in a Changing Climate* (2019): https://www.ipcc.ch/site/assets/uploads/sites/3/2019/11/06_SROCC_Ch02_FINAL.pdf.

15. M. Gilaberte-Búrdalo, F. López-Martín, M. R. Pino-Otín & J. I. López-Moreno, 'Impacts of climate change on ski industry', *Environmental Science & Policy* 44 (2014), 51–61.

16. L. G. Thompson, H. H. Brecher, E. Mosley-Thompson, D. R. Hardy & B. G. Mark, 'Glacier loss on Kilimanjaro continues unabated', *Proceedings of the National Academy of Sciences of the USA* 106 (2009), 19770.

17. P. Wester, A. Mishra, A. Mukherji & A. B. Shrestha (eds), *The Hindu Kush Himalaya Assessment: Mountains, Climate Change, Sustainability and People* (Springer, 2019): https://link.springer.com/book/10.1007/978-3-319-92288-1.

REFERENCES

18. M. Oppenheimer *et al.*, 'Sea Level Rise and Implications for Low-lying Islands, Coasts and Communities' in *IPCC 2019: Special Report on the Ocean and Cryosphere in a Changing Climate* (2019): https://www.ipcc.ch/srocc/chapter/chapter-4-sea-level-rise-and-implications-for-low-lying-islands-coasts-and-communities/.

19. J. Holder, N. Kommenda & J. Watts, 'The three-degree world: the cities that will be drowned by global warming', *Guardian* (3 November 2017): https://www.theguardian.com/cities/ng-interactive/2017/nov/03/three-degree-world-cities-drowned-global-warming.

20. M. Maslin, *Climate Change: A Very Short Introduction* (Oxford University Press, 2021).

21. Md. N. Islam & A. van Amstel, *Bangladesh I: Climate Change Impacts, Mitigation and Adaptation in Developing Countries* (Springer Nature, 2018).

22. Environment Agency, *Thames Estuary 2100: Managing Flood Risk Through London and the Thames Estuary* (2012): https://assets.publishing.service.gov.uk/government/uploads/system/uploads/attachment_data/file/322061/LIT7540_43858f.pdf.

23. R. M. DeConto & D. Pollard, 'Contribution of Antarctica to past and future sea-level rise', *Nature* 531 (2016), 591–7.

24. P. J. Sousounis & C. M. Little, *Climate Change Impacts on Extreme Weather* (Air Worldwide Corp., 2017): https://www.air-worldwide.com/SiteAssets/Publications/White-Papers/documents/Climate-Change-Impacts-on-Extreme-Weather.

25. T. Knutson *et al.*, 'Tropical cyclones and climate change assessment: Part II: Projected response to anthropogenic warming', *Bulletin of the American Meteorological Society* 101 (2019), E303–22.

26. J. B. Elsner, J. P. Kossin & T. H. Jagger, 'The increasing intensity of the strongest tropical cyclones', *Nature* 455 (2008), 92–5.

27. Y. Y. Loo, L. Billa & A. Singh, 'Effect of climate change on seasonal monsoon in Asia and its impact on the variability of monsoon rainfall in Southeast Asia', *Geoscience Frontiers* 6 (2015), 817–23.

28. WMO, *Integrated Flood Management Tools Series: Urban Flood Management in a Changing Climate* (2012): https://library.wmo.int/doc_num.php?explnum_id=7333.

29. R. A. Feely, S. C. Doney & S. R. Cooley, 'Ocean acidification: Present conditions and future changes in a high-CO_2 world', *Oceanography* 22 (2009), 36–47.

30. S. C. Doney, V. J. Fabry, R. A. Feely & J. A. Kleypas, 'Ocean acidification: The other CO_2 problem', *Annual Review of Marine Science* 1 (2009), 169–92.

31. C. Mbow et al., 'Food Security' in *Climate Change and Land: an IPCC Special Report on Climate Change, Desertification, Land Degradation, Sustainable Land Management, Food Security, and Greenhouse Gas Fluxes in Terrestrial Ecosystems* (2019): https://www.ipcc.ch/srccl/.

32. J. R. Porter et al., 'Food Security and Food Production Systems' in *Climate Change 2014: Impacts, Adaptation, and Vulnerability, Part A: Global and Sectoral Aspects. Contributions of Working Group II to the Fifth Assessment Report of the Intergovernmental Panel on Climate Change* (2014): https://www.ipcc.ch/site/assets/uploads/2018/02/WGIIAR5-Chap7_FINAL.pdf.

33. FAO, *The State of Food Security and Nutrition in the World 2019: Safeguarding Against Economic Slowdowns and Downturns* (2019): http://www.fao.org/3/ca5162en/ca5162en.pdf.

34. FAO, *Climate Change, Water and Food Security* (2008): http://www.fao.org/3/i2096e/i2096e.pdf.

35. N. Watts et al., 'The 2019 report of The *Lancet* Countdown on health and climate change: Ensuring that the health of a child born today is not defined by a changing climate', *Lancet* 394 (2019), 1836–78.

36. K. R. Smith et al., 'Human Health: Impacts, Adaptation, and Co-benefits' in *Climate Change 2014: Impacts, Adaptation, and Vulnerability, Part A: Global and Sectoral Aspects. Contribution of Working Group II to the Fifth Assessment Report of the Intergovernmental Panel on Climate Change* (2014): https://www.ipcc.ch/site/assets/uploads/2018/02/WGIIAR5-Chap11_FINAL.pdf.

37. IOM, *Migration and Climate Change* (2008).

38. D. J. Cantor, *Cross-border Displacement, Climate Change and Disasters: Latin America and the Caribbean* (UNHCR, 2018): https://www.unhcr.org/uk/protection/environment/5d4a7b737/cross-border-displacement-climate-change-disasters-latin-america-caribbean.html.

39. Q. Wodon, A. Liverani, G. Joseph & N. Bougnoux, *Climate Change and Migration: Evidence from the Middle East and North Africa* (World Bank, 2014).

40. A. Panda, 'Climate induced migration from Bangladesh to India: Issues and challenges', *SSRN Electronic Journal* 2010, 1–28.

41. EJF, *Beyond Borders: Our Changing Climate – Its Role in Conflict and Displacement* (2017): https://ejfoundation.org/resources/downloads/BeyondBorders.pdf.

42. IPCC, *Global Warming of 1.5°C: An IPCC Special Report* (2019): https://www.ipcc.ch/site/assets/uploads/sites/2/2019/06/SR15_Full_Report_High_Res.pdf.

43. IPCC, *Renewable Energy Sources and Climate Change Mitigation* (2012): https://www.ipcc.ch/site/assets/uploads/2018/03/SRREN_Full_Report-1.pdf.

44. G. Jia *et al.*, 'Land–Climate Interactions' in *Climate Change and Land: an IPCC Special Report on Climate Change, Desertification, Land Degradation, Sustainable Land Management, Food Security, and Greenhouse Gas Fluxes in Terrestrial Ecosystems* (2019): https://www.ipcc.ch/srccl/.

45. C40 Cities Climate Leadership Group, GCoM & IPCC, *Summary for Urban Policy Makers: What the IPCC Special Report on Global Warming of 1.5°C Means for Cities* (2018): https://www.c40.org/researches/summary-for-urban-policymakers-what-the-ipcc-special-report-on-global-warming-of-1-5-c-means-for-cities.

46. O. Lucon *et al.*, 'Buildings' in *Climate Change 2014: Mitigation of Climate Change. Contribution of Working Group III to the Fifth Assessment Report of the Intergovernmental Panel on Climate Change* (2014): https://www.ipcc.ch/site/assets/uploads/2018/02/ipcc_wg3_ar5_chapter9.pdf.

47. R. Sims *et al.*, 'Transport' in *Climate Change 2014: Mitigation of Climate Change. Contribution of Working Group III to the Fifth Assessment Report of the*

Intergovernmental Panel on Climate Change (2014): https://www.ipcc.ch/site/
assets/uploads/2018/02/ipcc_wg3_ar5_chapter8.pdf.

48. Transport & Environment, *Roadmap to Decarbonising European Aviation* (2018):
     https://www.transportenvironment.org/sites/te/files/publications/2018_10_
     Aviation_decarbonisation_paper_final.pdf.

49. J. Heywood, 'The virtues of meeting virtually in a time of climate crisis', *RSA* (11
     June 2019): https://www.thersa.org/discover/publications-and-articles/
     rsa-blogs/2019/06/the-virtues-of-meeting-virtually-in-a-time-of-climate-
     crisis.

50. B. A. Jones *et al.*, 'Zoonosis emergence linked to agricultural intensification and
     environmental change', *Proceedings of the National Academy of Sciences of
     the USA* 110 (2013), 8399–404.

51. RSA, *Creative Citizen, Creative State – The Principled and Pragmatic Case for a
     Universal Basic Income* (2015): https://www.thersa.org/globalassets/reports/
     rsa_basic_income_20151216.pdf.

52. E. O. Wilson, *Half-Earth: Our Planet's Fight for Life* (W. W. Norton & Co., 2016).

53. K. Gi, F. Sano, K. Akimoto, R. Hiwatari & K. Tobita, 'Potential contribution of
     fusion power generation to low-carbon development under the Paris
     Agreement and associated uncertainties', *Energy Strategy Reviews* 27
     (2020), 1–11.

CHAPTER 6: POWER OF THE INDIVIDUAL

1. M. H. Goldberg, S. van der Linden, E. Maibach & A. Leiserowitz, 'Discussing
    global warming leads to greater acceptance of climate science', *Proceedings
    of the National Academy of Sciences of the USA* 116 (2019), 14804–5.

2. Q. Schiermeier, 'Eat less meat: UN climate-change report calls for change to
    human diet', *Nature* 572 (2019), 291–2.

3. T. Raphaely & D. Marinova, 'Flexitarianism: Decarbonising through flexible
    vegetarianism', *Renewable Energy* 67 (2014), 90–96.

4. P. Scarborough *et al.*, 'Dietary greenhouse gas emissions of meat-eaters, fish-eaters, vegetarians and vegans in the UK', *Climatic Change* 125 (2014), 179–92.

5. W. Willett *et al.*, 'Food in the Anthropocene: the EAT–*Lancet* Commission on healthy diets from sustainable food systems', *Lancet* 393 (2019), 447–92.

6. IPCC, *Climate Change and Land: an IPCC Special Report on Climate Change, Desertification, Land Degradation, Sustainable Land Management, Food Security, and Greenhouse Gas Fluxes in Terrestrial Ecosystems* (2019): https://www.ipcc.ch/site/assets/uploads/2019/11/SRCCL-Full-Report-Compiled-191128.pdf.

7. E. Röös, T. Garnett, V. Watz & C. Sjörs, *The Role of Dairy and Plant Based Dairy Alternatives in Sustainable Diets* (Swedish University of Agricultural Sciences, 2018): https://pub.epsilon.slu.se/16016/1/roos_e_et_al_190304.pdf.

8. V. Roeben, 'The global community is finally acting on climate change, but we need to switch to renewable energy faster', *The Conversation* (22 August 2019): https://theconversation.com/the-global-community-is-finally-acting-on-climate-change-but-we-need-to-switch-to-renewable-energy-faster-119841.

9. IRENA, *Renewable Power Generation Costs in 2018* (2019): https://www.irena.org/-/media/Files/IRENA/Agency/Publication/2019/May/IRENA_Renewable-Power-Generations-Costs-in-2018.pdf.

10. A. Rowell, *Communities, Councils & a Low-Carbon Future: What We Can Do If Governments Won't* (Transition Books, 2010).

11. US Department of Energy, Energy Saver, *Tips on Saving Money and Energy in Your Home* (2017): https://www.energy.gov/sites/prod/files/2017/10/f37/Energy_Saver_Guide-2017-en.pdf.

12. Committee on Climate Change, *UK Housing: Fit for the Future?* (2019): https://www.theccc.org.uk/wp-content/uploads/2019/02/UK-housing-Fit-for-the-future-CCC-2019.pdf.

13. Committee on Climate Change, *Net Zero: The UK's Contribution to Stopping Global Warming* (2019): https://www.theccc.org.uk/wp-content/

uploads/2019/05/Net-Zero-The-UKs-contribution-to-stopping-global-warming.pdf.

14. Office for National Statistics, 'Road transport and air emissions' (2019): https://www.ons.gov.uk/economy/environmentalaccounts/articles/roadtransportandairemissions/2019-09-16.

15. A. Neves & C. Brand, 'Assessing the potential for carbon emissions savings from replacing short car trips with walking and cycling using a mixed GPS-travel diary approach', *Transportation Research Part A: Policy and Practice* 123 (2019), 130–46.

16. WHO, *Health in the Green Economy: Health Co-benefits of Climate Change Mitigation – Transport Sector* (2011): http://extranet.who.int/iris/restricted/bitstream/handle/10665/70913/9789241502917_eng.pdf;jsessionid=509F19AF7D4BAB84921BAAF102B713E4?sequence=1.

17. European Environment Agency, *The First and Last Mile – The Key to Sustainable Urban Transport: Transport and Environment Report 2019* (2020): https://www.eea.europa.eu//publications/the-first-and-last-mile.

18. L. Cozzi, 'Growing preference for SUVs challenges emissions reductions in passenger car market' (IEA, 2019): https://www.iea.org/commentaries/growing-preference-for-suvs-challenges-emissions-reductions-in-passenger-car-market.

19. N. Kommenda, 'How your flight emits as much CO_2 as many people do in a year', *Guardian* (19 July 2019): https://www.theguardian.com/environment/ng-interactive/2019/jul/19/carbon-calculator-how-taking-one-flight-emits-as-much-as-many-people-do-in-a-year.

20. BBC News, 'Climate change: Should you fly, drive or take the train?' (24 August 2019): https://www.bbc.co.uk/news/science-environment-49349566.

21. UNFCCC, Carbon Offset Platform, 'What is offsetting?' (2020): https://offset.climateneutralnow.org/aboutoffsetting.

22. P. Collinson, 'How to get your pension fund to divest from fossil fuels', *Guardian* (9 May 2015): https://www.theguardian.com/money/2015/may/09/how-get-pension-fund-divest-fossil-fuels (2015).

23. 350.org, Green Century Funds & Trillium Asset Management, *Make a Clean Break: Your Guide to Fossil Fuel Free Investing – An Updated Guide to Personal Divestment and Reinvestment* (2017): https://trilliuminvest.com/whitepapers/make-a-clean-break-your-guide-to-fossil-fuel-free-investing.

24. IEEFA, *The Financial Case for Fossil Fuel Divestment* (2018): http://ieefa.org/wp-content/uploads/2018/07/Divestment-from-Fossil-Fuels_The-Financial-Case_July-2018.pdf.

25. As You Sow, 'Carbon Clean 200™: Investing in a clean energy future – 2018 Q1 performance update' (2018): https://www.asyousow.org/report/clean200-2018-q1.

26. Friends of the Earth, 'Natural resources: Overconsumption and the environment' (2020): https://friendsoftheearth.uk/natural-resources.

27. S. Helm, J. Serido, S. Ahn, V. Ligon & S. Shim, 'Materialist values, financial and pro-environmental behaviors, and well-being', *Young Consumers* 20 (2019), 264–84.

28. WRAP, Recycle Now, 'Reduce waste' (2020): https://www.recyclenow.com/reduce-waste.

29. K. Williamson, A. Satre-Meloy, K. Velasco & K. Green, *Climate Change Needs Behavior Change: Making the Case for Behavioral Solutions to Reduce Global Warming* (Rare, 2018): https://rare.org/wp-content/uploads/2019/02/2018-CCNBC-Report.pdf.

30. Global Recycling Day, *Recycling: The Seventh Resource Manifesto* (2018): https://www.globalrecyclingday.com/wp-content/uploads/2017/12/ManifestoFINAL.pdf.

31. Circle Economy, *The Circularity Gap Report 2020* (2020): https://pacecircular.org/sites/default/files/2020-01/Circularity%20Gap%20Report%202020.pdf.

32. M. Berners-Lee, *How Bad Are Bananas? The Carbon Footprint of Everything* (Profile, 2010).

33. S. Laville, 'One year to save the planet: a simple, surprising guide to fighting the climate crisis in 2020', *Guardian* (7 January 2020): https://www.theguardian.com/environment/2020/jan/07/save-the-planet-guide-fighting-climate-crisis-veganism-flying-earth-emergency-action.

34. BBC News, 'What is Extinction Rebellion and what does it want?' (7 October 2019): https://www.bbc.co.uk/news/uk-48607989.

35. M. Taylor, J. Watts & J. Bartlett, 'Climate crisis: 6 million people join latest wave of global protests', *Guardian* (27 September 2019): https://www.theguardian.com/environment/2019/sep/27/climate-crisis-6-million-people-join-latest-wave-of-worldwide-protests.

36. E. Marris, 'Why young climate activists have captured the world's attention', *Nature* 573 (2019), 471–2.

37. M. Berners-Lee, *There Is No Planet B: A Handbook for the Make or Break Years* (Cambridge University Press, 2019).

CHAPTER 7: CORPORATE POSITIVE POWER

1. HowMuch.net, 'Charted: The companies making the most money in 2019' (2019): https://howmuch.net/articles/worlds-largest-companies-by-revenue.

2. McKinsey Global Institute, *Measuring the Economic Impact of Short-termism* (2017): https://www.mckinsey.com/~/media/McKinsey/Featured%20Insights/Long%20term%20Capitalism/Where%20companies%20with%20a%20long%20term%20view%20outperform%20their%20peers/MGI-Measuring-the-economic-impact-of-short-termism.ashx.

3. BSR, *Business in a Climate-Constrained World: Creating an Action Agenda for Private-sector Leadership on Climate Change* (2015): http://www.bsr.org/reports/bsr-bccw-creating-action-agenda-private-sector-leadership-climate-change.pdf.

4. CDP, 'World's top green businesses revealed in the CDP A List' (2019): https://www.cdp.net/fr/articles/companies/worlds-top-green-businesses-revealed-in-the-cdp-a-list.

5. CDP, *Climate Action and Profitability: CDP S&P 500 Climate Change Report 2014* (2014): https://6fefcbb86e61af1b2fc4-c70d8ead6ced550b4d987d7c03fcdd1d.ssl.cf3.rackcdn.com/cms/reports/documents/000/000/845/original/CDP-SP500-leaders-report-2014.pdf?1472032950.

6. Microsoft, 'Microsoft will be carbon negative by 2030' (2020): https://blogs. microsoft.com/blog/2020/01/16/microsoft-will-be-carbon-negative-by-2030/.

7. Sky, Sky Zero, 'We're going net zero carbon by 2030. Because the world can't wait' (2020): https://www.skygroup.sky/sky-zero.

8. BP, 'BP sets ambition for net zero by 2050, fundamentally changing organisation to deliver' (2020): https://www.bp.com/en/global/corporate/ news-and-insights/press-releases/bernard-looney-announces-new- ambition-for-bp.html.

9. D. A. Lubin, T. Nixon & C. Mangieri, *Transparency: The Pathway to Leadership for Carbon Intensive Businesses* (Reuters *et al.*, 2019): https://www.reuters.com/ media-campaign/brandfeatures/transparency-report/the-pathway-to- leadership-for-carbon-intensive-businesses-feb2019.pdf.

10. Natural Capital Partners, *The CarbonNeutral Protocol: The Global Standard for Carbon Neutral Programmes* (2020): https://carbonneutral.com/pdfs/The_ CarbonNeutral_Protocol_Jan_2020.pdf.

11. UN Global Compact, *Guide to Corporate Sustainability: Shaping a Sustainable Future* (2014): https://www.globalcompact.de/wAssets/docs/Nachhaltigkeits- CSR-Management/un_global_compact_guide_to_corporate_sustainability.pdf.

12. CDP, *Mind the Science* (2015): https://sciencebasedtargets.org/.

13. Science Based Targets, *Science-Based Target Setting Manual* (2019): https:// sciencebasedtargets.org/wp-content/uploads/2017/04/SBTi-manual.pdf.

14. TCFD, *Recommendations of the Task Force on Climate-related Financial Disclosures* (2017): https://www.fsb-tcfd.org/wp-content/uploads/2017/06/ FINAL-2017-TCFD-Report-11052018.pdf.

15. T. Bruckner *et al.*, 'Energy Systems' in *Climate Change 2014: Mitigation of Climate Change. Contribution of Working Group III to the Fifth Assessment Report of the Intergovernmental Panel on Climate Change* (2014): https://www. ipcc.ch/site/assets/uploads/2018/02/ipcc_wg3_ar5_chapter7.pdf.

16. The Climate Group, *Smarter Energy Use: Businesses Doing More with Less: EP100 Progress and Insights Report* (2019): https://www.theclimategroup.org/energy.

17. World Green Building Council, *Doing Right by Planet and People: The Business Case for Health and Wellbeing in Green Building* (2018): https://www.worldgbc.org/sites/default/files/WorldGBC%20-%20Doing%20Right%20by%20Planet%20and%20People%20-%20April%202018_0.pdf.

18. IRENA, *Corporate Sourcing of Renewable Energy: Market and Industry Trends – REmade Index* (2018): https://irena.org/-/media/Files/IRENA/Agency/Publication/2018/May/IRENA_Corporate_sourcing_2018.pdf.

19. Natural Capital Partners, *Green Gas Certificates: Help Companies Report Lower Scope 1 Emissions* (2018): https://assets.naturalcapitalpartners.com/downloads/Green_Gas_Certificate_Factsheet.pdf.

20. The Climate Group, *Business Driving Demand for Electric Vehicles: EV100 Progress and Insights Annual Report 2019* (2019): https://www.theclimategroup.org/about_re100.

21. BSR, *Transitioning to Low-Carbon Fuel: A Business Guide for Sustainable Trucking in North America – A Working Paper from the Future of Fuels Working Group* (2014): http://www.bsr.org/reports/BSR_Future_of_Fuels_Transitioning_to_Low_Carbon_Fuel.pdf.

22. Ellen MacArthur Foundation, *Towards the Circular Economy: Economic and Business Rationale for an Accelerated Transition* (2013): https://www.ellenmacarthurfoundation.org/assets/downloads/publications/Ellen-MacArthur-Foundation-Towards-the-Circular-Economy-vol.1.pdf.

23. Circle Economy, *The Circularity Gap Report 2020* (2020): https://pacecircular.org/sites/default/files/2020-01/Circularity%20Gap%20Report%202020.pdf.

24. Ellen MacArthur Foundation, *Growth Within: A Circular Economy Vision for a Competitive Europe* (2015): https://www.ellenmacarthurfoundation.org/assets/downloads/publications/EllenMacArthurFoundation_Growth-Within_July15.pdf.

25. European Union, *Moving Towards a Circular Economy with EMAS: Best Practices to Implement Circular Economy Strategies (with Case Study Examples)* (2017): https://ec.europa.eu/environment/emas/pdf/other/report_EMAS_Circular_Economy.pdf.

26. C. B. Bhattacharya, S. Sen & D. Korschun, 'Using corporate social responsibility to win the war for talent', *MIT Sloan Management Review* 49 (2008), 37–44.

27. Global Tolerance, *The Values Revolution* (2015): http://crnavigator.com/materialy/bazadok/405.pdf.

28. R. G. Eccles, K. Miller Perkins & G. Serafeim, 'How to become a sustainable company', *MIT Sloan Management Review* 53 (2012), 43–50.

29. ISO, *Introduction to ISO 14001: 2015* (2015): https://www.iso.org/files/live/sites/isoorg/files/store/en/PUB100371.pdf.

30. J. M. Sullivan, *Creating Employee Champions: How to Drive Business Success Through Sustainability Engagement Training* (Routledge, 2014).

31. P. Polman & C. B. Bhattacharya, 'Engaging employees to create a sustainable business', *Stanford Social Innovation Review* Fall 2016, 34–9.

32. J. B. Rodell, J. E. Booth, J. W. Lynch & K. P. Zipay, 'Corporate volunteering climate: Mobilizing employee passion for societal causes and inspiring future charitable action', *Academy of Management Journal* 60 (2017), 1662–81.

33. EY and UN Global Compact, *The State of Sustainable Supply Chains: Building Responsible and Resilient Supply Chains* (2016): https://www.ey.com/Publication/vwLUAssets/EY-building-responsible-and-resilient-supply-chains/$FILE/EY-building-responsible-and-resilient-supply-chains.pdf.

34. CDP, *Missing Link: Harnessing the Power of Purchasing for a Sustainable Future* (2017): https://www.bsr.org/reports/Report-Supply-Chain-Climate-Change-2017.pdf.

35. UN Global Compact, *Supply Chain Sustainability: A Practical Guide for Continuous Improvement* (2010): https://www.bsr.org/reports/BSR_UNGC_SupplyChainReport.pdf.

36. R. Isaksson, P. Johansson & K. Fischer, 'Detecting supply chain innovation potential for sustainable development', *Journal of Business Ethics* 97 (2010), 425–42.

REFERENCES

37. OECD, *Is There a Role for Blockchain in Responsible Supply Chains?* (2019): https://mneguidelines.oecd.org/Is-there-a-role-for-blockchain-in-responsible-supply-chains.pdf.

38. C. A. Adams, 'Sustainability and the Company of the Future' in *Reinventing the Company for the Digital Age* (BBVA, 2014), pp. 413–30.

39. Ellen MacArthur Foundation, *Completing the Picture: How the Circular Economy Tackles Climate Change* (2019): https://www.ellenmacarthurfoundation.org/assets/downloads/Completing_The_Picture_How_The_Circular_Economy-_Tackles_Climate_Change_V3_26_September.pdf.

40. BSR, *Redefining Sustainable Business: Management for a Rapidly Changing World* (2018): https://www.bsr.org/reports/BSR_Redefining_Sustainable_Business.pdf.

41. Corporate Leaders Group, 'More than 200 leading businesses urge UK Government to deliver resilient recovery plan' (2020): https://www.corporateleadersgroup.com/reports-evidence-and-insights/news-items/leading-businesses-urge-uk-government-to-deliver-resilient-recovery-plan.

42. UN Global Compact, *The Ambition Loop: How Business and Government Can Advance Policies that Fast Track Zero-carbon Economic Growth* (2018): https://www.unglobalcompact.org/library/5648.

43. UN Global Compact, *Guide for Responsible Corporate Engagement in Climate Policy: A Caring for Climate Report* (2013): https://www.unglobalcompact.org/docs/issues_doc/Environment/climate/Guide_Responsible_Corporate_Engagement_Climate_Policy.pdf.

44. AccountAbility and UN Global Compact, *Towards Responsible Lobbying: Leadership and Public Policy* (2005): https://www.unglobalcompact.org/docs/news_events/8.1/rl_final.pdf.

45. L. Georgeson & M. Maslin, 'Estimating the scale of the US green economy within the global context', *Palgrave Communications* 5 (2019), 1–12.

46. UNFCCC, *Energizing Entrepreneurs to Tackle Climate Change: Addressing Climate Change Through Innovation* (2018): https://unfccc.int/ttclear/misc_/StaticFiles/gnwoerk_static/brief12/bd80d2dd55e64d8ebdbc07752108c52c/af75fb524aa042e2a4f795ba6f29196f.pdf.

CHAPTER 8: GOVERNMENT SOLUTIONS

1. E. Somanathan *et al.*, 'National and Sub-national Policies and Institutions' in *Climate Change 2014: Mitigation of Climate Change. Contribution of Working Group III to the Fifth Assessment Report of the Intergovernmental Panel on Climate Change* (2014): https://www.ipcc.ch/site/assets/uploads/2018/02/ipcc_wg3_ar5_chapter15.pdf.

2. C. C. Jaeger, K. Hasselmann, G. Leipold, D. Mangalagiu & J. D. Tàbara (eds), *Reframing the Problem of Climate Change: From Zero Sum Game to Win–Win Solutions* (Earthscan, 2012).

3. PwC, *Innovation: Government's Many Roles in Fostering Innovation* (2010): https://www.pwc.com/gx/en/technology/pdf/how-governments-foster-innovation.pdf.

4. T. Bruckner *et al.*, 'Energy Systems' in *Climate Change 2014: Mitigation of Climate Change. Contribution of Working Group III to the Fifth Assessment Report of the Intergovernmental Panel on Climate Change* (2014): https://www.ipcc.ch/site/assets/uploads/2018/02/ipcc_wg3_ar5_chapter7.pdf.

5. UNEP, *Waste to Energy: Considerations for Informed Decision-Making* (2019): https://wedocs.unep.org/handle/20.500.11822/28413.

6. L. E. Erickson & M. Jennings, 'Energy, transportation, air quality, climate change, health nexus: Sustainable energy is good for our health', *AIMS Public Health* 4 (2017), 47–61.

7. R. Sims *et al.*, 'Transport' in *Climate Change 2014: Mitigation of Climate Change. Contribution of Working Group III to the Fifth Assessment Report of the Intergovernmental Panel on Climate Change* (2014): https://www.ipcc.ch/site/assets/uploads/2018/02/ipcc_wg3_ar5_chapter8.pdf.

8. D. Coady, I. Parry, N. Le & B. Shang, *Global Fossil Fuel Subsidies Remain Large: An Update Based on Country-Level Estimates* (IMF working paper, 2019): https://www.imf.org/en/Publications/WP/Issues/2019/05/02/Global-Fossil-Fuel-Subsidies-Remain-Large-An-Update-Based-on-Country-Level-Estimates-46509.

REFERENCES

9. IISD, *Fossil Fuel to Clean Energy Subsidy Swaps: How to pay for an energy revolution* (2019): https://www.iisd.org/sites/default/files/publications/fossil-fuel-clean-energy-subsidy-swap.pdf.

10. OECD, *Taxing Energy Use* (2019): https://www.oecd.org/tax/tax-policy/brochure-taxing-energy-use-2019.pdf.

11. World Bank, *Carbon Tax Guide: A Handbook for Policy Makers* (2017): https://www.cbd.int/financial/2017docs/wb-carbontaxguide2017.pdf.

12. World Green Building Council, *Bringing Embodied Carbon Upfront: Coordinated Action for the Building and Construction Sector to Tackle Embodied Carbon* (2019): https://www.worldgbc.org/embodied-carbon.

13. Better Buildings Partnership, *Low Carbon Retrofit Toolkit: A Roadmap to Success* (2010): http://www.betterbuildingspartnership.co.uk/sites/default/files/media/attachment/bbp-low-carbon-retrofit-toolkit.pdf.

14. J. Bogner et al., 'Waste Management' in *Climate Change 2007: Mitigation. Contribution of Working Group III to the Fourth Assessment Report of the Intergovernmental Panel on Climate Change* (2007): https://archive.ipcc.ch/publications_and_data/ar4/wg3/en/ch10.html.

15. IUCN, *Synergies Between Climate Mitigation and Adaptation in Forest Landscape Restoration* (2015): https://portals.iucn.org/library/node/45203.

16. M. Löf, P. Madsen, M. Metslaid, J. Witzell & D. F. Jacobs, 'Restoring forests: Regeneration and ecosystem function for the future', *New Forests* 50 (2019), 139–51.

17. FAO, *Sustainable Agriculture for Biodiversity: Biodiversity for Sustainable Agriculture* (2018): http://www.fao.org/3/a-i6602e.pdf.

18. Y. Cerqueira et al., 'Ecosystem Services: The Opportunities of Rewilding in Europe' in *Rewilding European Landscapes* (Springer, 2015), pp. 47–64.

19. N. Pettorelli et al., 'Making rewilding fit for policy', *Journal of Applied Ecology* 55 (2018), 1114–25.

20. OECD, *Towards Sustainable Land Use: Aligning Biodiversity, Climate and Food Policies* (2020): http://www.oecd.org/environment/resources/towards-sustainable-land-use-aligning-biodiversity-climate-and-food-policies.pdf.

21. Committee on Climate Change, *Land Use: Policies for a Net Zero UK* (2020):

https://www.theccc.org.uk/wp-content/uploads/2020/01/Land-use-Policies-for-a-Net-Zero-UK.pdf.

22. Compassion in World Farming, *Turning the Food System Round: The Role of Government in Evolving to a Food System That Is Nourishing, Sustainable, Equitable and Humane* (2019): https://www.ciwf.org.uk/media/7436369/how-to-transition-to-a-nourishing-sustainable-equitable-and-humane-food-system-2019.pdf.

23. A. R. Camilleri, R. P. Larrick, S. Hossain & D. Patino-Echeverri, 'Consumers underestimate the emissions associated with food but are aided by labels', *Nature Climate Change* 9 (2019), 53–8.

24. EPHA & HCWH Europe, *How Can the EU Farm to Fork Strategy Contribute? Public Procurement for Sustainable Food Environments* (2019): https://epha.org/wp-content/uploads/2019/12/public-procurement-for-sustainable-food-environments-epha-hcwh-12-19.pdf.

25. O. De Schutter, *The Power of Procurement: Public Purchasing in the Service of Realizing the Right to Food* (2014): https://www.pianoo.nl/sites/default/files/documents/documents/thepowerofprocurement.pdf.

26. Department of Health & Social Care, *Consultation on Restricting Promotions of Products High in Fat, Sugar and Salt by Location and By Price* (2019): https://assets.publishing.service.gov.uk/government/uploads/system/uploads/attachment_data/file/770704/consultation-on-restricting-price-promotions-of-HFSS-products.pdf.

27. True Animal Protein Price Coalition, *Aligning Food Pricing Policies with the European Green Deal: True Pricing of Meat and Dairy in Europe, Including CO_2 Costs* (2020): https://www.tappcoalition.eu/true-pricing-of-food.

28. L. Cornelsen & A. Carriedo, *Health-related Taxes on Food and Beverages* (Food Research Collaboration, 2015): https://foodresearch.org.uk/publications/health-related-taxes-on-food-and-beverages/.

29. Defra, *Clean Air Strategy 2019* (2019): https://assets.publishing.service.gov.uk/government/uploads/system/uploads/attachment_data/file/770715/clean-air-strategy-2019.pdf.

REFERENCES

30. WWF, *Saving the Earth: A Sustainable Future for Soils and Water* (2018): https://www.wwf.org.uk/sites/default/files/2018-04/WWF_Saving_The_Earth_Report_HiRes_DPS_0.pdf.

31. Global Food Security & UK Water Partnership, *Agriculture's Impacts on Water Quality* (2015): https://www.foodsecurity.ac.uk/publications/archive/page/5/.

32. X. Poux & P. M. Aubert, *An Agroecological Europe in 2050: Multifunctional Agriculture for Healthy Eating. Findings from the Ten Years for Agroecology (TYFA) Modelling Exercise* (IDDRI, 2018): https://www.iddri.org/sites/default/files/PDF/Publications/Catalogue%20Iddri/Etude/201809-ST0918EN-tyfa.pdf.

33. UNDP, *Taxes on Pesticides and Chemical Fertilizers* (2017): https://www.undp.org/content/dam/sdfinance/doc/Taxes%20on%20pesticides%20and%20chemical%20fertilizers%20_%20UNDP.pdf.

34. PMR & ICAP, *Emissions Trading in Practice: A Handbook on Design and Implementation* (2016): https://icapcarbonaction.com/en/?option=com_attach&task=download&id=364.

35. European Environment Agency, *National Climate Change Vulnerability and Risk Assessments in Europe, 2018* (2018): https://www.eea.europa.eu/publications/national-climate-change-vulnerability-2018.

36. UNESCO and UNFCCC, *Action for Climate Empowerment: Guidelines for Accelerating Solutions Through Education, Training and Public Awareness* (2016): https://unfccc.int/sites/default/files/action_for_climate_empowerment_guidelines.pdf.

37. OECD, *Integrating Climate Change Adaptation into Development Co-operation: Policy Guidance* (2009): https://www.oecd.org/env/cc/44887764.pdf.

38. OECD, *Climate-resilient Infrastructure: Policy Perspectives* (2018): http://www.oecd.org/environment/cc/policy-perspectives-climate-resilient-infrastructure.pdf.

39. J. Rydge, M. Jacobs & I. Granoff, *Ensuring New Infrastructure Is Climate-Smart* (New Climate Economy, 2015): https://newclimateeconomy.report/2015/wp-content/uploads/sites/3/2014/08/Ensuring-infrastructure-is-climate-smart.pdf.

40. S. L. Lewis & M. A. Maslin, *The Human Planet: How We Created the Anthropocene* (Penguin, 2018).

41. RSA, *Creative Citizen, Creative State – The Principled and Pragmatic Case for a Universal Basic Income* (2015): https://www.thersa.org/globalassets/reports/rsa_basic_income_20151216.pdf.

CHAPTER 9: SAVING OUR PLANET AND OURSELVES

1. C. C. Jaeger, K. Hasselmann, G. Leipold, D. Mangalagiu & J. D. Tàbara (eds), *Reframing the Problem of Climate Change: From Zero Sum Game to Win–Win Solutions* (Earthscan, 2012).

2. C. I. Bradford & J. F. Linn, *Global Governance Reform: Breaking the Stalemate* (Brookings Institution Press, 2007).

3. D. P. Rapkin, J. R. Strand & M. W. Trevathan, 'Representation and governance in international organizations', *Politics and Governance* 4 (2016), 77–89.

4. D. Tapscott, 'A Bretton Woods for the 21st century', *Harvard Business Review* (2014): https://hbr.org/2014/03/a-bretton-woods-for-the-21st-century.

5. Bretton Woods Project, *What Are the Main Criticisms of the World Bank and IMF?* (2019): https://www.brettonwoodsproject.org/wp-content/uploads/2019/06/Common-Criticisms-FINAL.pdf.

6. N. McCarthy, 'Oil and gas giants spend millions lobbying to block climate change policies', *Forbes* (25 March 2019): https://www.forbes.com/sites/niallmccarthy/2019/03/25/oil-and-gas-giants-spend-millions-lobbying-to-block-climate-change-policies-infographic/#1bddde527c4f.

7. K. Das, H. van Asselt, S. Droege & M. Mehling, *Making the International Trade System Work for Climate Change: Assessing the Options* (Climate Strategies, 2018): https://climatestrategies.org/wp-content/uploads/2018/07/CS-Report-_Trade-WP4.pdf.

8. UNCTAD, *Trade and Development Report 2018: Power, Platforms and the Free Trade Delusion* (2018): https://unctad.org/en/PublicationsLibrary/tdr2018_en.pdf.

9. J. Hickel, *The Divide: A Brief Guide to Global Inequality and Its Solutions* (Heinemann, 2017).

10. D. Kar, *Financial Flows and Tax Havens: Combining to Limit the Lives of Billions of People* (Norwegian School of Economics *et al.*, 2016): https://gfintegrity.org/report/financial-flows-and-tax-havens-combining-to-limit-the-lives-of-billions-of-people/.

11. United Nations, 'Security Council: Current members' (2020): https://www.un.org/securitycouncil/content/current-members.

12. L. Cabrera, *Strengthening Security, Justice, and Democracy Globally: The Case for a United Nations Parliamentary Assembly* (Commission on Global Security, Justice & Governance, 2015): https://www.stimson.org/wp-content/files/Commission_BP_Cabrera.pdf.

13. Wikipedia, 'Democracy Index' (2020): https://en.wikipedia.org/wiki/Democracy_Index.

14. V. Popovski, *Reforming and Innovating the United Nations Security Council* (Commission on Global Security, Justice & Governance, 2015): https://www.stimson.org/wp-content/files/Commission_BP_Popovski1.pdf.

15. L. Georgeson, 'Why is there still no World Environment Organisation?' *The Conversation* (22 April 2014): https://theconversation.com/why-is-there-still-no-world-environment-organisation-25792.

16. United Nations System, 'Total revenue by agency' (2018): https://www.un.org/annualreport/index.html.

17. J. D. Ostry, P. Loungani & D. Furceri, *Neoliberalism: Oversold?* (IMF, 2016): https://www.imf.org/external/pubs/ft/fandd/2016/06/pdf/ostry.pdf.

18. A. Kapoor & B. Debroy, 'GDP is not a measure of human well-being', *Harvard Business Review* (4 October 2019): https://hbr.org/2019/10/gdp-is-not-a-measure-of-human-well-being.

19. J. A. McGregor & N. Pouw, 'Towards an economics of well-being', *Cambridge Journal of Economics* 41 (2016), 1123–42.

20. Credit Suisse Research Institute, *Global Wealth Report 2019* (2019): https://www.credit-suisse.com/about-us-news/en/articles/media-releases/

global-wealth-report-2019--global-wealth-rises-by-2-6--driven-by-201910.
html.

21. N. Maxwell, 'From Knowledge to Wisdom: The Need for an Academic Revolution', *London Review of Education* 5 (2007), 97–115.

22. United Nations, 'Sustainability Development Goals' (2020): https://sustainabledevelopment.un.org/.

23. Editorial, 'Get the Sustainable Development Goals back on track', *Nature* 577 (2020), 7–8.

24. Wikipedia, 'List of treaties unsigned or unratified by the United States' (2020): https://en.wikipedia.org/wiki/List_of_treaties_unsigned_or_unratified_by_the_United_States.

25. Credit Suisse Research Institute, *Asia in Transition* (2018): https://www.credit-suisse.com/about-us-news/en/articles/news-and-expertise/emerging-asia-will-produce-more-than-half-of-global-output-201811.html.

26. World Bank, 'GDP, PPP (current international $) – European Union, United States, China' (2020): https://data.worldbank.org/indicator/NY.GDP.MKTP.PP.CD?end=2018&locations=EU-US-CN&start=1960.

27. United Nations, *World Population Prospects 2019: Highlights* (2019): https://population.un.org/wpp/Publications/Files/WPP2019_Highlights.pdf.

28. US National Intelligence Council, *Global Trends 2030: Alternative Worlds* (2012): https://globaltrends2030.files.wordpress.com/2012/11/global-trends-2030-november2012.pdf.

29. H. Kharas, *The Unprecedented Expansion of the Global Middle Class: An Update* (Brookings Institution, 2017): https://www.brookings.edu/wp-content/uploads/2017/02/global_20170228_global-middle-class.pdf.

30. Friends of the Earth, 'Natural resources: Overconsumption and the environment' (2020): https://friendsoftheearth.uk/natural-resources.

31. McKinsey Global Institute, *Asia's Future Is Now* (2019): https://www.mckinsey.com/~/media/McKinsey/Featured%20Insights/Asia%20Pacific/Asias%20future%20is%20now/Asias-future-is-now-final.ashx.

32. IPCC, *Global Warming of 1.5°C: An IPCC Special Report* (2019): https://www.ipcc.ch/site/assets/uploads/sites/2/2019/06/SR15_Full_Report_High_Res.pdf.

33. FAO, *Zero-deforestation Commitments: A New Avenue Towards Enhanced Forest Governance?* (2018): http://www.fao.org/3/i9927en/I9927EN.pdf.

34. UNEP, *Sustainable Consumption and Production: A Handbook for Policymakers* (2015): https://sustainabledevelopment.un.org/content/documents/1951Sustainable%20Consumption.pdf.

35. UNCTAD, *Growth and Poverty Eradication: Why Addressing Inequality Matters* (2013): https://unctad.org/en/PublicationsLibrary/presspb2013d4_en.pdf.

36. RSA, *Creative Citizen, Creative State – The Principled and Pragmatic Case for a Universal Basic Income* (2015): https://www.thersa.org/globalassets/reports/rsa_basic_income_20151216.pdf.

37. United Nations, *The Sustainable Development Goals Report 2019* (2019): https://unstats.un.org/sdgs/report/2019/The-Sustainable-Development-Goals-Report-2019.pdf.

38. N. Stern *et al.*, *Stern Review: The Economics of Climate Change* (UK Government, 2006).

39. R. Stavins *et al.*, 'International Cooperation: Agreements & Instruments' in *Climate Change 2014: Mitigation of Climate Change. Contribution of Working Group III to the Fifth Assessment Report of the Intergovernmental Panel on Climate Change* (2014): https://www.ipcc.ch/site/assets/uploads/2018/02/ipcc_wg3_ar5_chapter13.pdf.

40. UN Division for Sustainable Development, *A Guidebook to the Green Economy* (2012): https://sustainabledevelopment.un.org/content/documents/738GE%20Publication.pdf.

41. K. Williamson, A. Satre-Meloy, K. Velasco & K. Green, *Climate Change Needs Behavior Change: Making the Case for Behavioral Solutions to Reduce Global Warming* (Rare, 2018): https://rare.org/wp-content/uploads/2019/02/2018-CCNBC-Report.pdf.

42. S. Laville, 'One year to save the planet: a simple, surprising guide to fighting the climate crisis in 2020', *Guardian* (7 January 2020): https://www.theguardian.com/environment/2020/jan/07/save-the-planet-guide-fighting-climate-crisis-veganism-flying-earth-emergency-action.

43. E. Somanathan *et al.*, 'National and Sub-national Policies and Institutions' in *Climate Change 2014: Mitigation of Climate Change. Contribution of Working Group III to the Fifth Assessment Report of the Intergovernmental Panel on Climate Change* (2014): https://www.ipcc.ch/site/assets/uploads/2018/02/ipcc_wg3_ar5_chapter15.pdf.

44. D. Ivanova *et al.*, 'Quantifying the potential for climate change mitigation of consumption options', *Environmental Research Letters* (2020).

45. BSR, *Business in a Climate-Constrained World: Creating an Action Agenda for Private-sector Leadership on Climate Change* (2015): http://www.bsr.org/reports/bsr-bccw-creating-action-agenda-private-sector-leadership-climate-change.pdf.

46. AccountAbility and UN Global Compact, *Towards Responsible Lobbying: Leadership and Public Policy* (2005): https://www.unglobalcompact.org/docs/news_events/8.1/rl_final.pdf.

47. UNFPA, *Unfinished Business: The Pursuit of Rights and Choices for All* (2019): https://www.unfpa.org/sites/default/files/pub-pdf/UNFPA_PUB_2019_EN_State_of_World_Population.pdf.

AFTERWORD

1. P. Friedlingstein, M. O'Sullivan, M. W. Jones, *et al.*, 'Global Carbon Budget 2020', Earth System Science Data, 12, 3269–3340 (2020): https://doi.org/10.5194/essd-12-3269-2020.

EXTRA FACTS ABOUT OUR PLANET: EARTH

Our planet.

Our home.

Our blue marble floating in dark, empty space.

The third planet from the Sun.[1]

Earth goes around the Sun every 365.256 days, travelling at 107,200 km/h.[2]

The equator is 40,075 km long.[2]

It would take you 8,015 hours or 668 days to walk the length of it (if you could).

The surface area of the Earth[3] is 510 million km².

Half of the surface area is in the Tropics.

The volume of the Earth[2] is 1083 billion km^3.

The mass of the Earth[2] is 600 billion tonnes.

Earth is the densest planet in the Solar System.[1]

29% of the Earth's surface is land – from rainforests to arid sandy deserts.[3]

The highest peak on land is Mount Everest, at 8,848 m above sea level.[4]

The lowest point on land is the Dead Sea, in 2020 the surface was 434 m below sea level.[5]

71% of the Earth's surface is water – oceans, seas, lakes and rivers.[3]

The deepest part of the ocean is the Challenger Deep in the Mariana Trench, at 10,984 m below sea level.[6]

The mean surface temperature is 14°C.[7]

The coldest recorded temperature is -89.2°C.[8]

The warmest recorded temperature is 56.7°C.[8]

The largest ocean is the Pacific Ocean, covering a third of the Earth's surface (165,250,000 km^2).[9]

The four longest rivers in the world are:

Nile (6,693 km),

Amazon (6,436 km),

Yangtze (6,300 km),

Mississippi (6,275 km).[10]

The largest delta is the Ganges Delta at over 100,000 km^2.[11]

The five largest deserts in the world are:

Antarctica (14 million km^2),

Sahara (8.6 million km^2),

Arabian (2.3 million km^2),

Gobi (1.3 million km^2),

Kalahari (0.9 million km^2).[12]

The three largest rainforests are:

Amazon (5,500,000 km²),

Congolese (1,780,000km²),

New Guinea (288,000 km²).[13]

We have discovered 1.9 million different species alive today, and there could be as many as 1 trillion species on Earth.[14, 15]

There are over 7.8 billion people on Earth – all of whom are unique.[16]

And just one of you!

REFERENCES

1. NASA, 'Our Solar System' (2009): https://www.jpl.nasa.gov/edu/pdfs/ss-high.pdf.
2. NASA, 'Earth fact sheet' (2020): https://nssdc.gsfc.nasa.gov/planetary/factsheet/earthfact.html.
3. M. Pidwirny, 'Introduction to the oceans', *Fundamentals of Physical Geography* (2nd edn, 2006): http://www.physicalgeography.net/fundamentals/8o.html.
4. BBC News, 'Nepal and China agree on Mount Everest's height' (8 April 2010): http://news.bbc.co.uk/1/hi/world/south_asia/8608913.stm.
5. Israel Oceanographic & Limnological Research, 'Long-term changes in the Dead Sea' (2020): https://isramar.ocean.org.il/isramar2009/DeadSea/LongTerm.aspx.

6.   J. V. Gardner, A. A. Armstrong, B. R. Calder & J. Beaudoin, 'So, how deep *is* the Mariana Trench?', *Marine Geodesy* 37 (2014), 1–13.

7.   NASA Earth Observatory, 'World of Change: Global temperatures' (2020): https://earthobservatory.nasa.gov/world-of-change/decadaltemp.php.

8.   Arizona State University, 'World Meteorological Organization Global Weather & Climate Extremes Archive' (2020): https://wmo.asu.edu/content/world-meteorological-organization-global-weather-climate-extremes-archive.

9.   J. A. Quinn & S. L. Woodward, *Earth's Landscape: An Encyclopedia of the World's Geographic Features* (ABC-CLIO, 2015).

10. J. Boenigk, S. Wodniok & E. Glücksman, *Biodiversity and Earth History* (Springer, 2015), doi:10.1007/978-3-662-46394-9.

11. R. Ramachandran *et al.*, 'Integrated Management of the Ganges Delta, India' in *Coasts and Estuaries: The Future*, ed. E. Wolanski, J. W. Day, M. Elliott & R. Ramachandran (Elsevier, 2019).

12. C. Holzapfel, 'Deserts' in *Encyclopedia of Ecology* (2nd edn), ed. B. Fath (Elsevier, 2008), doi:10.1016/B978-0-444-63768-0.00326-7, pp. 447–66.

13. FAO, *The State of the World's Forests: Forest Pathways to Sustainable Development* (2018): http://www.fao.org/3/I9535EN/i9535en.pdf.

14. A. D. Chapman, *Numbers of Living Species in Australia and the World* (2009): https://www.environment.gov.au/system/files/pages/2ee3f4a1-f130-465b-9c7a-79373680a067/files/nlsaw-2nd-complete.pdf.

15. K. J. Locey & J. T. Lennon, 'Scaling laws predict global microbial diversity', *Proceedings of the National Academy of Sciences of the USA* 113 (2016), 5970–75.

16. United Nations, *World Population Prospects 2019: Highlights* (2019): https://population.un.org/wpp/Publications/Files/WPP2019_Highlights.pdf.

FURTHER READING

HISTORY OF OUR PLANET AND SPECIES

Charlesworth, B., and D. Charlesworth, *Evolution: A Very Short Introduction* (Oxford University Press, 2017)

Christakis, N. A., *Blueprint: The Evolutionary Origins of a Good Society* (Little, Brown Spark, 2019)

Humphrey, L., and C. Stringer, *Our Human Story – Illustrated* (Natural History Museum, 2018)

Langmuir, C. H., and W. Broecker, *How to Build a Habitable Planet: The Story of Earth from the Big Bang to Humankind* (Princeton University Press, 2012)

Lenton, T., *Earth System Science: A Very Short Introduction* (Oxford University Press, 2016)

Lenton, T., and A. Watson, *Revolutions That Made the Earth* (Oxford University Press, 2013)

Lewis, S. L., and M. A. Maslin, *The Human Planet: How We Created the Anthropocene* (Penguin and Yale University Press, 2018)

Maslin, M. A., *The Cradle of Humanity* (Oxford University Press, 2017)

McMichael, A. J., *Climate Change and the Health of Nations: Famines, Fevers and the Fate of Populations* (Oxford University Press, 2019)

Ruddiman, W. F., *Plows, Plagues, and Petroleum: How Humans Took Control of Climate* (Princeton Science Library, 2016)

Zalasiewicz, J., and M. Williams, *The Goldilocks Planet: The Four Billion Year Story of Earth's Climate* (Oxford University Press, 2012)

SCIENCE OF CLIMATE CHANGE

Archer, D., *Global Warming: Understanding the Forecast*, 2nd edn (John Wiley, 2011)

Dessler, A. E., *The Science and Politics of Global Climate Change: A Guide to the Debate*, 3rd edn (Cambridge University Press, 2019)

Emanuel, K., *What We Know About Climate Change*, updated edition (MIT Press, 2018)

Houghton, J. T., *Global Warming: The Complete Briefing*, 5th edn (Cambridge University Press, 2015)

IPCC, *Climate Change 2021: The Physical Science Basis. Contribution of Working Group I to the Sixth Assessment Report of the Intergovernmental Panel on Climate Change* (2021)

Maslin, M. A., *Climate Change: A Very Short Introduction* (Oxford University Press, 2021)

Maslin, M. A., and S. Randalls (eds), *Future Climate Change: Critical Concepts in the Environment* (4 volumes containing reproductions of 85 of the most important papers published on climate change), Routledge Major Work Collection (Routledge, 2012)

National Climate Assessment, *Volume I: Climate Science Special Report* (2018); https://science2017.globalchange.gov/

Romm, J., *Climate Change: What Everyone Needs to Know* (Oxford University Press, 2018)

POLITICS AND HISTORY OF CLIMATE CHANGE

Corfee-Morlot, J., *et al.*, 'Climate science in the public sphere', *Philosophical Transactions of the Royal Society A* 365/1860 (2007), 2741–76

Frankopan, P., *The New Silk Roads: The Present and Future of the World* (Bloomsbury, 2019)

Giddens, A., *The Politics of Climate Change*, 2nd edn (Polity Press, 2011)

Gupta, J., *The History of Global Climate Governance* (Cambridge University Press, 2014)

Klein, N., *On Fire: The Burning Case for a Green New Deal* (Allen Lane, 2019)

Leggett, J. K., *The Winning of the Carbon War: Power and Politics on the Front Lines of Climate and Clean Energy* (Crux Publishing, 2018)

Mann, M., *The New Climate War: The Fight to Take Back Our Planet* (PublicAffairs, 2021)

Maslin, M., 'The five corrupt pillars of climate change denial', *The Conversation* (28 November 2019); https://theconversation.com/the-five-corrupt-pillars-of-climate-change-denial-122893

Maslin, M., 'Five climate change science misconceptions – debunked', *The Conversation* (15 September 2019); https://theconversation.com/five-climate-change-science-misconceptions-debunked-122570

Metcalf, G. E., *Paying for Pollution: Why a Carbon Tax Is Good for America* (Oxford University Press, 2019)

Meyer, A., *Contraction and Convergence: The Global Solution to Climate Change* (Green Books, 2015)

Oreskes, N., and E. M. Conway, *Merchants of Doubt: How a Handful of Scientists Obscured the Truth on Issues from Tobacco Smoke to Global Warming* (Bloomsbury, 2012)

Thunberg, G., *No One Is Too Small to Make a Difference* (Penguin, 2019)

Weart, S. R., *The Discovery of Global Warming*, New Histories of Science, Technology, and Medicine (Harvard University Press, 2008)

IMPACTS OF CLIMATE CHANGE

Bardgett, R., *Earth Matters: How Soil Underlies Civilization* (Oxford University Press, 2016)

Costello, A., *et al.*, 'Managing the health effects of climate change', *Lancet* 373 (2009), 1693–733

Friel, S., *Climate Change and the People's Health* (Oxford University Press, 2019)

Garcia, R. A., *et al.*, 'Multiple dimensions of climate change and their implications for biodiversity', *Science* 344 (2014), 486–96

IPCC, *Climate Change 2021: Impacts, Adaptation, and Vulnerability. Contribution of Working Group II to the Sixth Assessment Report of the Intergovernmental Panel on Climate Change* (2021)

Lynas, M., *Our Final Warning: Six Degrees of Climate Emergency* (Fourth Estate, 2020)

National Climate Assessment, *Volume II: Impacts, Risks, and Adaptation in the United States* (2018); https://nca2018.globalchange.gov/

Stern, N., *The Economics of Climate Change: The Stern Review* (Cambridge University Press, 2007)

Wallace-Wells, D., *The Uninhabitable Earth: A Story of the Future* (Penguin, 2019)

Watts, N., *et al.*, 'The 2020 report of The *Lancet* Countdown on Health and Climate Change', *Lancet* [396] (2020): 1129–306: https:/www.thelancet.com/countdown-health-climate

GLOBAL POLITICS, ECONOMICS AND GOVERNANCE

Berners-Lee, M., *There Is No Planet B: A Handbook for the Make or Break Years* (Cambridge University Press, 2019)

Grubb, M., *Planetary Economics: Energy, Climate Change and the Three Domains of Sustainable Development* (Routledge, 2014)

IPCC, *Climate Change 2022: Mitigation of Climate Change. Contribution of Working Group III to the Sixth Assessment Report of the Intergovernmental Panel on Climate Change* (2022)

Jackson, T., *Prosperity Without Growth: Economics for a Finite Planet* (Routledge, 2016)

Mazzucato, M., *The Value of Everything: Making and Taking in the Global Economy* (Penguin, 2019)

Raworth, K., *Doughnut Economics: Seven Ways to Think Like a 21st-century Economist* (Random House, 2017)

Royal Society, *People and the Planet*, Royal Society Science Policy Centre Report 01/12 (2012), p. 81

Sachs, J., *The Ages of Globalization* (Columbia University Press, 2020)

21ST-CENTURY SOLUTIONS

Behrens, P., T. Bosker and D. Ehrhardt (eds), *Food and Sustainability* (Oxford University Press, 2020)

Buck, H. J., *After Geoengineering: Climate Tragedy, Repair, and Restoration* (Verso, 2019)

Centre for Alternative Technology, *Zero Carbon Britain: Rising to the Climate Emergency* (CAT Publications, 2019); https://www.cat.org.uk/new-report-zero-carbon-britain-rising-to-the-climate-emergency/

Cole, L., *Who Cares Wins: Reasons for Optimism in Our Changing World* (Penguin, 2020)

Figueres, C., and T. Rivett-Carnac, *The Future We Choose: Surviving the Climate Crisis* (Manilla Press, 2020)

Flannery, T., *Atmosphere of Hope: Solutions to the Climate Crisis* (Penguin, 2015)

Georgeson, L., M. Poessinouw and M. Maslin, 'Assessing the definition and measurement of the global green economy', *Geo: Geography and Environment* 4(1), ISSN 2054-4049, doi: 10.1002/geo2.36 (2017)

Goodall, C., *What We Need to Do Now: For a Zero Carbon Future* (Profile Books, 2020)

Hawken, P., *Drawdown: The Most Comprehensive Plan Ever Proposed to Reverse Global Warming* (Penguin, 2018)

Hayhoe, K., *The Answer to Climate Change: And Why We Can Have Hope* (Atria/One Signal Publishers, 2021)

Helm, D., *Net Zero: How We Stop Causing Climate Change* (William Collins, 2020)

IPCC, *Climate Change 2021: Impacts, Adaptation, and Vulnerability. Contribution of Working Group II to the Sixth Assessment Report of the Intergovernmental Panel on Climate Change* (2021)

Morton, O., *The Planet Remade: How Geoengineering Could Change the World* (Granta, 2016)

Roaf, S., *et al.*, *Adapting Buildings and Cities for Climate Change* (Routledge, 2009)

ACKNOWLEDGEMENTS

The author would like to thank the following people:

Johanna, Alexandra and Abbie for surviving lockdown together as a family bubble and allowing me to write this book.

My agent Tessa David, who grasped my mad idea when I pitched it and helped me refine the style you see in this unique book.

My editors at Penguin Emily Robertson and Susannah Bennett, my fact checker Carole Roberts and copy-editor Emma Horton.

All the wonderful staff at University College London, Rezatec Ltd and Trove Research. A special thanks to the deep insights of Siva Niranjan and Simon L. Lewis.

Thanks to all my brilliant dedicated colleagues in climatology, chemistry, economics, engineering, geology, geography, history, the humanities, social science, economics, medicine, the arts and many other subjects who continue to strive to understand, predict and mitigate our negative impacts on our planet.

And finally, thanks to Sun Tzu, whoever 'he' or 'they' were, for writing *The Art of War* over 2,500 years ago and inspiring me to write this book.

INDEX

INDEX